AF699719

INNOVATION LEADERS

Copyright © Infinite Ideas Limited, 2008

The right of Innovaro Ltd to be identified as the author of this book has been asserted in accordance with the Copyright, Designs and Patents Act 1988.

First published in 2008 by
Infinite Ideas Limited
36 St Giles
Oxford, OX1 3LD
United Kingdom
www.infideas.com

All rights reserved. Except for the quotation of small passages for the purposes of criticism or review, no part of this publication may be reproduced, stored in a retrieval system or transmitted in any form or by any means, electronic, mechanical, photocopying, recording, scanning or otherwise, except under the terms of the Copyright, Designs and Patents Act 1988 or under the terms of a licence issued by the Copyright Licensing Agency Ltd, 90 Tottenham Court Road, London W1T 4LP, UK, without the permission in writing of the publisher. Requests to the publisher should be addressed to the Permissions Department, Infinite Ideas Limited, 36 St Giles, Oxford, OX1 3LD, UK, or faxed to +44 (0)1865 514777.

A CIP catalogue record for this book is available from the British Library

ISBN 978–1–904902–95–9

Brand and product names are trademarks or registered trademarks of their respective owners.

Research by Innovaro
Edited by Caroline Jones
Designed and typeset by Julie Clements
Printed in Great Britain by Acorn Press, Swindon

INNOVATION LEADERS

CONTENTS

INTRODUCTION

PART 1 - INNOVATION PROFILES

PART 2 - ADDITIONAL INSIGHTS

CONCLUSION

INTRODUCTION

INNOVATION AND GROWTH

Innovation is the topic for twenty-first century businesses. It is now firmly on the corporate agenda and is seen by companies, analysts, governments and consumers as a primary source of competitive differentiation. As media interest has increasingly aligned around popular stories ranging from Apple and Google through to 3M and Sony, it has been hard to miss that innovation has been increasingly seen as one of the most important drivers of growth. Whether at a business unit, company or even country level, from FTSE 100 and Fortune 500 companies through to EU Framework programmes, the World Bank and the UN, innovation is increasingly being embraced as the source of improved performance, sustained growth and long-term socio-economic benefit.

Everyone certainly wants to gain from innovation and, as they join in the maelstrom of associated activities, many are seeking different perspectives of what innovation is and could be. Although the high level message is hugely attractive, many organisations, disciplines and even individuals have differing views on innovation:

- To some it is all about product launches and improving market share;
- To others it is more about the financial impacts such as revenue and margin growth;
- For some innovation is seen in a more technical manner and is often linked to patents and other tangible outputs;
- For others it is more about creativity, intellectual capital and the influence of organisational culture.
- Others simply see it as a means of improving their return on investment that, if well executed, delivers higher growth than alternative options.

Whichever of these has greatest individual resonance, across all these perspectives, the underlying commonality is that innovation per se is seen as being linked to positive outcomes. Whether financial, cultural or reputation enhancing, innovation is a highly attractive proposition: Innovation leaders, those companies that are able to better understand emerging opportunities, access and exploit new technologies, develop successful new products and services or even change their business models, are the corporate heroes of today.

As Wall Street analysts and company management place ever more demanding expectation and growth targets on organisations, innovation is seen as the most successful route for delivering organic growth. This book is about the companies that are today's innovation heroes, the organisations that CEOs want to head up, people want to work for and others seek to emulate. It is designed to provide in-depth insights into the leaders in each sector, the contexts within which they are innovating and the approaches that they are using to be at the top of their league.

INNOVATION LEADERS

The annual Innovation Leaders analysis identifies the most effective innovator in each of 25 major sectors. It is increasingly recognised as the most authoritative assessment of innovation performance around and is gradually extending to cover more and more companies and sectors. It differs significantly from other innovation assessments such as those used by Business Week, the EU and the OECD in that it is based on looking at which companies are the most effective innovators, not necessarily the largest. Rather than just looking at who simply spends most on innovation, who produces the most patents, who launches the most new products or whose name is at the front of mind of a group of CEOs, the Innovation Leaders analysis does not take a singular view. It uses assessment of innovation efficiency and effectiveness to highlight the companies that, although they may not be shouting about their innovation credentials, are actually getting on and delivering innovation and gaining direct impact to the top and bottom line. The Innovation Leaders based on performance in 2007/8 are a formidable group of companies that are leading change across their respective sectors:

Sector	Innovation Leader	HQ Country
Aerospace	Boeing	US
Airlines	Virgin Atlantic	UK
Automotive	BMW	Germany
Chemicals	BASF	Germany
Consumer Electronics	Samsung	South Korea
Energy	Shell	UK / Netherlands
Fashion Retail	H&M	Sweden
Food and Drink	PepsiCo	US
General Retail	Tesco	UK
Hotels and Leisure	Starwood	US
Household Goods	Reckitt Benckiser	UK
IT Hardware	Apple	US
IT Services	Infosys	India
Insurance	Aviva	UK
Logistics	UPS	US
Media / Entertainment	Google	US
Medical Devices	Medtronic	US
Office Equipment	Canon	Japan
Pharmaceuticals	Lilly	US
Retail Banking	Handelsbanken	Sweden
Software	Microsoft	US
Sportswear	Adidas	Germany
Telecom Equipment	Nokia	Finland
Telecom Operators	NTT DoCoMo	Japan
Toys and Games	LEGO	Denmark

Alongside these identified Innovation Leaders, in each sector there are a number of other companies that are demonstrating potential for future impact. Some of these may well be short-term contenders for the Innovation Leader recognition whilst others are more long-term bets – companies that are doing interesting things, changing the sector dynamics and using innovation in new and different ways. Some of these companies have been strong innovators in the past and are again getting focus while others are new organisations that are rising fast up the innovation hierarchy. The ones that we are watching are:

Sector	Ones we are Watching	
Aerospace	EADS	Embraer
Airlines	Qatar Airways	Net-Jets
Automotive	GM	Tata
Chemicals	Dow	DSM
Consumer Electronics	LG	Sony
Energy	GE	BP
Fashion Retail	Baugur	Giordano
Food and Drink	Nestle	Cargill
General Retail	Wal-Mart	WuMart
Hotels and Leisure	Hyatt	Jumeirah
Household Goods	P&G	Kao
IT Hardware	Logitech	Lenovo
IT Services	IBM	Accenture
Insurance	Munich Re	Allianz Group
Logistics	Fed-Ex	TNT
Media / Entertainment	News Corporation	Disney
Medical Devices	Smith and Nephew	Johnson & Johnson
Office Equipment	Brother	Xerox
Pharmaceuticals	Genentech	Merck
Retail Banking	Barclays	Zopa
Software	Adobe	Salesforce.com
Sportswear	Nike	Puma
Telecom Equipment	Sony Ericsson	Qualcomm
Telecom Operators	Vodafone	Korea Telecom
Toys and Games	Nintendo	LeapFrog

38.69
17.13
41.47
27.09
22.47
+0.07
+0.13
+0.46

IMPACT ON SHARE PRICES

In the world of innovation, there has been an assumed linkage between innovation success and sustained growth. Over the years varied academics in the US and Europe have found correlations between innovation activity and corporate growth, while numerous government departments have mapped share price growth of high investors in research and development against FTSE, NASDAQ and Dow Jones indexes. Add into the mix OECD, World Bank, KPMG and PWC analysis and the guiding view for those involved in innovation has been that, as a means to generate successful, share price-influencing growth, it is largely more successful and more sustainable than a purely merger and acquisition strategy.

Over the past eight years, as the Innovation Leaders analysis has been undertaken and updated each year, we have had a unique set of companies to assess and track in terms of revenue and margin growth as well as share price rise or fall. The results have been quite revealing: half of the companies identified as Innovation Leaders in their respective sectors back in January 2005 subsequently demonstrated an increase in share price of at least 50% over the succeeding two years. In addition, the top five performers over this period all had growth in share price of over 60%. Between 1st January 2005 and 1st January 2007:

- Apple's share price increased by 163%;
- Google's share price increased by 139%;
- Rolls-Royce's share price increased by 89%;
- Toyota's share price increased by 64%; and
- Canon's share price increased by 61%.

Over the same period, the Dow went up by 15%, NASDAQ by 20% and the FTSE 100 by 9%. As a portfolio, the Innovation Leaders shares have in fact exceeded the performance of all the major indexes for the past seven years.

In the past twelve months, the rise of the Innovation Leaders shares have continued apace: Between 1st January 2007 and 1st January 2008, the 2006/7 Innovation Leaders overall portfolio went up by 12% while the Dow only managed 7%, NASDAQ achieved even lower growth of 4% with the FTSE 100 bringing up the rear with a growth of only 3%. The top 5 Innovation Leaders performers over this period were:

- Apple, whose share price increased by another 135%;
- Nokia, whose share price increased by 91%;
- Google, whose share price increased by 52%;
- Adidas, whose share price increased by 36%; and
- Reckitt Benckiser, whose share price increased by 25%.

Innovation success, as demonstrated by the Innovation Leadership analysis is thus clearly not only directly related to share price growth, but, by comparison to all major indexes, also enables companies to outperform their peers.

RESEARCH METHODOLOGY

Assessing innovation performance accurately is a challenge that many have tried to meet. Some have attempted to look solely at one particular element of the innovation portfolio, for example brand value, R&D investment or new product introductions, while others have tried to gain a cross-sector view. The problem with both of these approaches is that the flavours and importance of the different innovation ingredients often vary significantly from one sector to the next. While R&D investment and patent activity are good indicators of innovation performance in, say, the pharmaceuticals sector, the world of retail innovation is all about new product launches, average margin and increased revenue per store.

Equally, underneath such top-level views there are a number of internal innovation drivers such as culture, organisational structure and investment, which all underpin long-term innovation capability. In order to gain a deep, insightful and validated perspective on who are the true innovation leaders on a sector-by-sector basis, the annual Innovation Leadership analysis assesses the performance of 1,500 of the world's top companies against eight key parameters.

The individual innovation and growth performance of the top companies within each sector are compared against each other and additional research around the differences between the leading five companies is undertaken to ensure that the perspectives gained are fully up to date. From this, the Innovation Leader in each of 25 major sectors is identified and profiled. These profiles are the basis of this book and are widely used to inform opinion.

The eight key areas that we research and input into the assessments are:

1. Organisational culture and supporting structure;

2. Strategic focus on innovation and its role in driving corporate growth;

3. Number of major new product launches and relative success ratios;

4. Growth in revenues, profits and market capitalisation;

5. Average revenue and margin per product or customer;

6. Investment in innovation-related activities such as R&D and marketing;

7. Brand value and human capital growth; and

8. Peer review from within the sector.

In addition, where appropriate, we look at recent intellectual property performance focused on US patents and trademarks since these are the best independent objective guide in this area. Together these are used in sector-specific algorithms which calculate a ratio of innovation effectiveness – essentially identifying which companies gain the most innovation impact in relationship to the investments they make. This provides what we believe to be the most accurate picture of the companies which are the most effective innovators in each sector, and allows us to highlight the respective key competitive strengths.

PART 1 - INNOVATION PROFILES

- Adidas
- Apple
- Aviva
- BASF
- BMW
- Boeing
- Canon
- Google
- Handelsbanken
- H&M
- Infosys
- LEGO
- Lilly
- Medtronic
- Microsoft
- Nokia
- NTT DoCoMo
- PepsiCo
- Reckitt Benckiser
- Samsung
- Shell
- Starwood
- Tesco
- UPS
- Virgin Atlantic
- Common Ingredients

Integrating consumer-evident technology innovation in all key products to enhance performance credentials

ADIDAS

Adidas is on the move and always has been: It has had an adventurous history since it first grew out of a family business in Herzogenaurach, Germany in the 1920s. With the hostile separation of two brothers' interests in the 1940s, nearly going bust in the 1980s and then executing two rescue operations, first by sending production offshore to Asia and then by reinventing itself into a design and marketing company, Adidas has riden the waves of change in the sports goods sector both up and down. Alongside its own brands, it owned the Saloman ski and sportswear brand for nearly a decade and now includes the Reebok, Taylormade Golf and Rockport brands in its stable.

Things are now definitely on track and, if the current marketing slogan, "Impossible is Nothing", is anything to go by, the company is brimming with confidence. This is not surprising when you consider that Adidas now consistently out-performs the rest of the sector and has enjoyed eight years of consecutive double digit net income growth. It is now the world's number two sports apparel manufacturer with total sales for 2007 of €10.3bn and profit growth of 9%.

Innovation Scorecard	
Innovation Culture	8
Strategic Focus	9
New Products	8
Managed Growth	9
Product Margin	8
Investment in Innovation	8
Innovation Brand Impact	9
Innovation Peer Review	8

Adidas-Salomon Key Data 2007	
Total Sales	€10.3bn
Gross Profit	€14.8bn
Profit Growth	9%
Gross Margin	47.4%
Net Income	€551m
R&D Spend	€84m
Brand Value	$4.7bn
Sales / Employee	€329k

Walk along any high street and it is clear that wearing sports clothing is definitely a fashion statement, and possibly an indication of athletic prowess. Adidas recognised this trend early on and has developed high-performance sports lines in collaboration with the likes of Stella McCartney, Yohji Yamamoto, Porsche Design and Rolland Berry. That said, the company does not sacrifice its commitment to improving sporting performance and aims to launch at least one major new technology or technological evolution per year. Even more than its peers, Adidas has put performance at the heart of its product portfolio and invests specifically to support this. R&D projects involve collaborations with professional and amateur athletes including Zinedine Zidane, Michael Ballack and Allyson Felix. In addition, Adidas works with clubs such as AC Milan and Bayern Munich to test and optimise products. Over recent years this has led to development of technologies such as ForMotion, which supports the core adistar and Supernova families, as well as the next generation of the Response and BOUNCE running shoes.

Alongside providing performance products, Adidas recognises that consumers make purchase decisions based not only on brand but also on availability, convenience and breadth of product offering. As a result the company has been refining its distribution proposition, concentrating on expanding its own outlets or 'controlled' space and improving retail relationships. There are now over 1000 Adidas stores around the world and, in the run up to the Beijing Olympics the company opened an average of two stores a month in China. By 2010, the aim is to generate at least 30% of the group's revenues through controlled space.

To keep its brand in the public focus Adidas has also sponsored sportsmen and women for many years. In 2008, 295 footballers, 64 rugby players, 71 tennis players, 24 basketball players and 8 golfers all benefited from its three stripe logo. One of the first prominent endorsers of Adidas equipment was American running legend Jesse Owens, the gold medalist at the 1936 Summer Olympics. As well as sponsoring the Beijing Olympics Adidas is also supporting the 2012 Olympic Games in London in a deal worth around $200 million: in this sector such brand awareness is critical. Going forward, Adidas is also embracing a number of niche sports and lifestyle activities fuelled by the current trends in individualism, health and wellbeing with, for example, a new woman's Yoga range. As it aims to take over world number 1 position in the sector from Nike, Adidas will be innovating incessantly both within existing sports and outside the core.

SECTOR OVERVIEW

Sportswear is very big business: In Europe alone, almost 90% of people under 35, and 76% of the population as a whole, buy at least one item of sportswear a year. Consumers benefit from a huge range of choice so brand awareness and credibility is pretty much everything. Small wonder that marketing budgets are enormous and performance enhancing technology considered the holy grail of any sports manufacturer. In addition, trends in health and well-being are acting as catalysts for change as governments and sports associations recognise the cost benefits of securing fitter, healthier nations. The range of activities that we participate in is also growing as action sports, such as snowboarding, extreme biking and in-line skating, catch up with traditional games such as soccer and basketball. This is not solely a Western phenomenon. Thanks to greater promotion, the massive fan-bases that support the top teams and the worldwide desire for a healthier lifestyle there is significant and growing markets in Asia. Overall, annual growth in sportswear manufacturing is expected to continue at an average of 5% over the next decade.

INNOVATION DRIVERS

One of the hottest races at the Beijing Olympics is not among the athletes, but the companies that outfit them. The gold medal is, of course, domination of the Chinese market. Sports in China were once seen as a luxury but, as the middle class has grown and the culture has changed, interest has exploded. The likes of Nike and Adidas, which have had a low key presence in the market for decades, compete for brand preference and are using the 2008 Olympics as their battlefield. Sponsorship is a major weapon for both companies. Nike, for example, sponsors 22 of the 28 competing Chinese federations and one of China's most promising athletes, Liu Xiang, wears the trademark swoosh.

But brand awareness is only part of the strategy – performance, and the perception of it, is also of major importance in this market. To keep ahead, the major players in the sportswear industry need to demonstrate innovation across the portfolio - in technology, product design, functionality, graphics and presentation. The psychological effect of dynamic, cushioned track shoes or drag-reducing textiles on running and swimming suits cannot be overlooked. Moreover, as a casual glance at MTV demonstrates, nor can the influence of fashion.

As aditional technology is embedded within products from Speedo's Fastskin swimsuits to the Nike + iPod cobination, new developments are currently blurring the lines between fashion and well-being. Look out for more clothing that can 'think' for itself by using IT connectivity to PCs and gym equipment, and yet more smarter fabrics which are able to respond to changes in the environment by adjusting their pore size or thickness at a given moment to facilitate moisture transmission.

Lastly, like other sectors, sportswear is also going green. Alongside a major recent reorganisation of the supply chain for this sector to make it more sustainable, there are also a number of product changes in play: Look out for Adidas' Grun, a collection of organic lifestyle apparel. With three guiding elements of 'Made From', 'Recycled' and 'Reground', Grun offers a range of products, including classic silhouettes like the Forum basketball sneaker and the ZX 500 runner, all made from recycled and natural materials. If it can work for Nike with its Considered line then this could also be big news for Adidas.

ONES WE ARE WATCHING

Nike

Nike, the world's largest sportswear company, is in growth mode. 2007/8 revenues of $13.5bn were up 13% on the year before while gross margin increased by 16%. As it seeks to incorporate yet higher performance credentials into its products, this mass-market player is pushing harder into the swimwear and soccer areas. The rise of Adidas seems to have acted as a catalyst for Nike which has started linking social networking to its stores and won a series of prizes for new products including the Revolutionary Support Sports Bra, the Considered 2K5 shoe and the Nike+ Air Zoom Moire. Nike is aiming at $23bn revenues by 2011.

Puma

The outcome of the brothers' split at Adidas and so also based in Herzogenaurach, Puma was, for a long time, a relatively small player in the sportswear sector. However, it gained significant growth and rising profitability in the early part of the decade when its products were increasingly worn by celebrities such as Madonna. Now, with a string of new golf, running and soccer products hitting the market, Puma is a strong player from both the fashion and performance perspectives. It has double digit revenue growth and has had particular recent success at the Rugby World Cup and at the African Cup of Nations where it equipped over half the teams.

Iconic, design-led, continual evolution of mutually reinforcing product, service and business model innovation

APPLE

Ask anyone to name an innovative company and odds-on they will mention Apple. It has had a reputation for leading-edge products since college dropouts Steve Jobs and Steve Wozniak first founded the company. Apple memorably unveiled the Macintosh in 1984 and proceeded to build a hard-core following of customers who appreciated good design and were willing to pay premium prices while overlooking any interoperability issues with Windows - a factor that Apple has subsequently addressed with its OS-X operating system. Amongst a wealth of innovations introduced over the years, the radical, all-in-one design of the iMac was one turning point, bringing colour and light to the drab world of computing. The same lateral thinking and attention to detail has been applied to products such as the original iPod in 2001, the PowerBook in 2003, Mac Mini in 2005 and, in 2007, the iPhone.

Innovation Scorecard

Innovation Scorecard	
Innovation Culture	9
Strategic Innovation Focus	10
New Products	10
Managed Growth	10
Product Margin	9
Investment in Innovation	9
Innovation Brand Impact	10
Innovation Peer Review	9

Apple Key Data 2007

Apple Key Data 2007	
Total Sales	$24.0bn
Sales Growth	24%
Net Income	$3.5bn
Net Income Growth	76%
Gross Margin	34%
R&D Spend	$782m
R&D Intensity	3.3%
Total Assets	$25.4bn

In its traditional computing market, Apple continues to maintain its position as a leading innovator, strengthening its range of high quality laptop and desktop products. This has led to total Mac unit sales growing to over 7m and delivering $10.3bn of core revenues in 2007 - an increase of 40% on 2006. Allied to this, the iPod range has been extended with the new iPod Touch and an increased capacity for the classic models. Despite the average unit price decreasing from $195 to $161, there was an increase in unit sales from 39.4m to 51.6m, and overall iPod sales generated $8.3bn.

The launch of the iPhone created global publicity and cemented Apple's reputation as a company that continuously produces truly innovative, sector-disrupting products. With sales in the fiscal year of 1.4m units, the iPhone contributed only a small proportion of Apple revenues. However it created huge consumer pull. Additionally the iPhone demonstrated Apple's ability to break through existing paradigms – it is sold at a fixed price, is linked to a specific contract and is only available through specific mobile operators. As with the introduction of iTunes, Apple has entered a new market and challenged existing business models. This is a key element of Apple's approach to innovation: while many see the product as the focus, for Apple it is the interplay between the product and the service that allows it to fundamentally change the underlying business model operating in the sector. First it was music and now it is movies. Via its iTunes service, Apple has challenged the way the music industry operates and the company has now expanded its music videos, audio-books, television shows, movies and podcasts reportoire still further with major links to all the key studios. In 2007 iTunes had its three billionth download and this helped music related products and services account for a significant $2.5bn of sales. Of course customers also expect a suitably 'Apple' experience when they enter a store. And they get it. Apple now has 197 stores worldwide and this has redefined computer retail.

All of the above innovations have translated into corresponding financial success with Apple's share price doubling in 2007. Total sales increased by 24% to just over $24bn while net income increased by a staggering 76% to just under $3.5bn. This is all ample evidence to highlight that innovation in Apple is not hype: Apple really delivers its goal to provide 'innovative integrated digital lifestyle solutions' and is a company that consistently demonstrates the power of innovation in driving profitable, sustained growth.

SECTOR OVERVIEW

Every few years the priorities in the IT sector seem to shift. Product developments have come thick and fast in the area of information technology. PDAs, disk drives, USB memory sticks, web-cams, laser printers, colour printers, scanners, digital cameras and DVD writers have all become part of everyday life for people around the world. Dramatic improvements in functionality have been enabled by the introduction of faster chips, simplified communication protocols, wireless technologies and network servers. The IT hardware sector is a maze of organisations playing in an ever interconnected space.

Firms like Intel, ARM, AMD and Infineon concentrate on chip development; niche operators including Creative, Logitech and Maxtor focus on peripherals; companies like Acer, Dell and Apple are broadening out from their PC heritage, and organisations including Fujitsu, HP and Toshiba, as well as several traditional consumer electronics companies such as Sony and Samsung, are involved in multiple arenas including workstations, laptops, printers, monitors, servers, cameras, projectors and an array of handheld devices including mobile phones which have become true mobile computers. All in all, this sector is now truly converging.

INNOVATION DRIVERS

The key drivers in this sector continue to be the increasing convergence of devices, and the continued trend towards the provision of a suite of web-based, lifestyle related products that allow for the easy inter-operability of music players, cameras, communications, and desktop applications. At the same time it hard to see the industry moving backwards in terms of the need to provide cool, good-looking and ultra-functional devices where product reliability and performance are hygiene factors. This will obviously play into the hands of the likes of Apple and Sony's Vaio business, but equally there will always be a requirement for solid, 'work-horse' machines for the corporate and personal market. This is where companies like Dell have succeeded in the past, and it is likely that further innovations will be made in the manufacture of low cost machines allied to the speedy delivery of units through the already renowned supply chain. Other drivers are the increased inter-operability between platforms such as the Windows and Mac operating systems. As many manufacturers chase the prize to be had from creating the leading convergent device, the need to offer plug and play operation across ubiquitous networks continues to be both a fundamental target and a major innovation challenge.

ONES WE ARE WATCHING

Logitech

Logitech, headquartered in Switzerland, is a world leader in personal peripherals and is continuing to broaden its product offering and its presence in the retail sector. This is fuelled by a trend among consumers to enhance their basic PC systems with more fully featured personal peripherals that add functionality and cordless freedom to their desktops. They are also purchasing supplementary devices designed for new applications and specific purposes such as gaming, multimedia, or audio and visual communication over the Internet. The company's products combine essential core technologies, continuing innovation, award-winning industrial design and excellent price performance.

Lenovo

Lenovo, the Chinese company, bought IBM's loss-making $10 billion PC business in 2005 with the intention of building a prominent global brand. The recent launch of the super-slim ThinkPad X300 is Lenovo's first bid for leadership in the world of laptops. Lenovo's stated goal is to continuously develop the most inventive and best-quality products with efficiency. The X300 is prominently featured at the Beijing Olympics, where Lenovo is one of the major sponsors. The company hopes this will lead to a renewed interest in the ThinkPad range which is one of the longest lasting brands in computing, now into its 16th year, and which has sold over 30 million.

Bundling technology partnerships into core products to change sector dynamics and consumer relationships

AVIVA

Aviva is the world's fifth-largest insurance group. It is one of the leading providers of life and pension products in Europe and is actively growing its long-term savings businesses in Asian markets, Australia and the USA. The group's recent history is one of mergers to gain scale and thus competitive advantage. Unable to compete on its own in a market where scale matters, General Accident merged with Commercial Union to form CGU in 1998, which itself merged with Norwich Union in 2000 to form CGNU plc. This then subsequently changed to Aviva in July 2002 to strengthen its brand name.

Aviva's main activities are organised into two primary strategic areas: long-term savings and fund management, which includes a range of long-term savings, investment and protection products in markets that offer significant opportunities for growth; and general insurance, health and related services which provide a broad range of competitive motor, property, health and related insurance services to individuals and small to medium-sized enterprises. In 2007, the company had premium income and investment sales of £38.6bn, and £364bn of assets under management. While most of the group's businesses have been rebranded Aviva, some of its strongest consumer brands remain, including Norwich Union.

Innovation Scorecard

Innovation Culture	6
Strategic Innovation Focus	7
New Products	7
Managed Growth	8
Product Margin	8
Investment in Innovation	8
Innovation Brand Impact	7
Innovation Peer Review	8

Aviva Key Data 2007

Total Sales	*£38.6bn*
Sales Growth	*25%*
Operating Profit	*£3.2bn*
Profit Growth	*1.1%*
Assets Under Management	*£364bn*
Employees	*59,000*
Sales / Employee	*£654k*
Savings Sales Growth	*25%*

Norwich Union is the UK's largest insurer, covering one in seven motor vehicles, and has a market share of around 15 per cent. It is one of the few insurers to embrace the latest telematic technologies to introduce more customised motor products and services for its customers, and has very much led the way in this area. Working in partnership with IBM and Orange, for the telematics software and network coverage respectively, 'Pay As You Drive' was trialed back in 2003 and launched across the UK in October 2006. It uses the latest GPS technology to allow premiums to be calculated based on when, where and how far customers drive their vehicles. The service works by installing a GPS device the size of a DVD case in the car. This stores information about each car journey before transmitting it automatically to Norwich Union via a secure GSM network. The journey data is then translated into a bill, similar to a mobile phone bill, which provides details of each element used to calculate the premium. This comprises a fixed monthly fee to cover risks such as fire and theft, comparable to a 'line rental' charge on a phone bill, and a variable amount based on recorded data such as mileage driven, roads used and time of day of each journey, etc.

For young drivers, the primary target market, rates peak at night between 11pm and 6am, when figures show that young motorists are at much greater risk of having a major accident and seriously injuring themselves or others. For drivers aged over 24 rates vary according to the time of day and where they drive, with off-peak rates falling outside the morning rush hour and midnight to 5am. A year after the launch customer feedback was overwhelmingly positive with 90% looking to renew their policy, and with 30% fewer claims reported amongst all "Pay As You Drive" insurance customers, the road safety benefit is immediately clear. In a sector traditionally focused on price-driven annual renewal, this new service has changed the frequency of customer interaction, deepened the relationship with Aviva and, of great concern to others in the sector, enabled Aviva to increase its share of the all important first-time drivers market. As this and similar technology-enabled, customer-impacting services are rolled out across the company's key markets, despite escalating price competition, future growth for Aviva is anticipated.

SECTOR OVERVIEW

Within the insurance sector there is scope to utilise the latest technology to introduce innovative ways of providing cover for customers. However the challenge for the insurance industry is to keep abreast of emerging changes and quickly respond by developing new products and services that are both relevant and cost-competitive. Whilst other segments of the financial services industry are moving ahead in terms of proactively scanning their environment and even appointing dedicated innovation executives, there is little evidence that the insurance market is changing in this direction. This is an oversight as the opportunities for those that do embrace a more innovative mindset, especially in the younger segments, will be enormous.

Some insurers are aiming to break out of more traditional ways of working by looking beyond technology enablement to developing different business models and different ways of presenting their products: Prudential, for example, offers no-claims discounts on home insurance in the same way that they are offered on motor insurance. In a bid to change consumer behaviours, the same company also provides discounts for health insurance tied to gym membership. 2007 was a challenging year for many in the financial services industry and this affected the insurance sector just as much as the large lenders. Time will tell whether this will have a positive effect on innovation as companies seek to recover lost profits and fuel future organic growth.

INNOVATION DRIVERS

Innovation is not something that is naturally associated with a sector which, after all, is run by actuaries. No surprise then that taking the entire financial services industry into consideration, insurance is the area that has seen least product innovation over the past decade. However, changing demographics, the increased levels of threat in some areas, the potential for using new technologies, and the desire to bundle some aspects of financial services provision is forcing some insurers to review their product types and the way they are delivered, and most importantly quoted. In addition changes in legislation around the world including the relaxing of pension regulations in the UK, will undoubtedly result in opportunities for more innovative pensions and insurance products. Moreover, as the developed world gradually becomes older and fatter, the savings and investment sectors as well as the core insurance sector need to produce products and services that match the needs of this changing demographic.

Similarly another major threat to the insurance sector- that of climate change and associated natural catastrophes - has resulted in a number of initiatives. In 2005 natural disasters killed 97,000 people and cost the insurance industry $83 billion. This has spurred the sector to launch initiatives such as Lloyd's 360 risk project to assess these risks. Swiss Re has introduced its Climate Adaptation Development Programme, which is designed to provide financial protection against weather risks in emerging countries, and the Climate Wise initiative backed by global leaders in the insurance sector aims to reduce risk associated with climate change. Linked to this, insuring risk from investments in Clean Development Mechanism approved projects that fail to deliver could add 4 to 5% of growth to insurance markets. However the focus in the insurance sector is clearly about better understanding risk rather than, just yet, offering any 'green' products.

Finally the effects of the global credit crunch and the associated economic slowdown will slowly work its way through the insurance sector and surely impact the way such companies innovate. We are already seeing some major players stopping withdrawals from some funds and it remains to be seen the extent to which this will increase, or decrease, innovative activity.

ONES WE ARE WATCHING

Munich Re
Munich Re Group is one of the world's largest re-insurers and the second-largest primary insurer in Germany. In common with other sectors the company has an Innovative Solutions Team which recognises trends, analyses markets and works across all divisions to come up with solutions for the future. In January 2008 Munich Re announced a record profit of €3.9bn for 2007 on the basis of key preliminary figures and is set to continue accruing benefit from continuously looking to the future and trying to understand how the world will change and why.

Allianz
The Allianz Group is one of the leading global services providers in insurance, banking and asset management. Innovation at Allianz is firmly embedded in the company's strategy: It is a key element in feeding the strategic pipeline for the long term and supporting the initiatives that are driving profitable growth. Allianz's group-wide innovation initiative, 'ideas to success', was launched in 2006 and generated over 28,000 ideas in its first year. Allianz is starting to see financial benefits from 'ideas to success' and are looking to bring innovation into all business areas and implant it into the DNA of the company.

Using strong market focus and innovation outside the immediate core areas of the business to drive growth

BASF

For such a large company with significant global impact, BASF keeps a remarkably low public profile. However, every time you slip on your 501s its worth remembering that one of the company's first main innovations was to develop a synthetic, low cost indigo dye which, in part, explains why the popular working clothes of many nations, and particularly the USA, are blue. BASF is, in fact, the world's leading chemical company and its portfolio ranges from chemicals, plastics, performance products and fine chemicals to crude oil and natural gas. As a long-term partner to most industries, BASF has grown to have 95,000 employees and, in 2007, posted sales of €57.95bn. The company is currently expanding its international activities with a particular focus on Asia. BASF has been a consistent performer with good margins, above-average revenue growth as well as high-net income growth.

BASF
The Chemical Company

Innovation Scorecard	
Innovation Culture	7
Strategic Innovation Focus	8
New Products	7
Managed Growth	8
Product Margin	8
Investment in Innovation	7
Innovation Brand Impact	6
Innovation Peer Review	7

BASF Key Data 2007	
Total Revenues	€57.95bn
Revenue Growth	10.2%
Earnings	€6.94bn
R&D Spend	€1.38bn
R&D Intensity	2.4%
Employees	95175
Revenue / Employee	€609k
R&D Spend / Patent	€4.7m

A core element of BASF's success has been its ability to identify key market requirements early on, recognise a core need and then develop an extraordinary range of products to address it. Allied to this, the company also understands and exploits diversification. For example, BASF was one of the first organisations to recognise the need to shift from bulk chemicals to higher value consumer focused products such as coatings, vitamins and magnetic tape for audio and video cassettes. Some see that it is this willingness to adapt which has ensured the company has remained ahead of its peers.

BASF's quest for innovation has always been along a path of complementary spaces: The company has grown in areas associated with its core markets, moving into animal nutrition as a result of its work in agricultural products and into petrochemicals because of its involvement in supporting exploration for oil and gas. BASF's core innovation model, or 'verbund', is one of several ingredients of sustained growth. It combines product, process, technology and business models. For instance, the 'production verbund' links manufacturing plants to create value-added chains of production where by-products and waste from one plant serves as the raw materials in others. Today, the verbund has become a cornerstone of the company's strategy and now extends outside the organisation to include BASF's business partners, customers and the communities in which the company operates.

The strategy is producing commercial results and industry recognition: In 2007, BASF was awarded the prestigious Design Plus award by the German Design Council for Ultrason, a high-temperature-resistant, transparent plastic now used in induction water boilers. In addition Toyota chose another plastic, Basotect, for its Lexus car hoods because it is lightweight, helps to reduce CO2 emissions and absorbs sound - thereby increasing driver comfort. BASF has also been making its own contribution to the sustainability agenda by producing plastics made from renewable sources and ionic liquids to improve solar cell performance. BASF is also changing its organisational structures to reinforce the search for innovation across different dimensions: in June of 2007, it restructured its business to be more aligned with customer-driven markets including the automotive and construction industries and created a broader 'BASF Care' group that encompasses health, personal care and nutrition. The strong performance over the years, stable investment in R&D and drive for looking beyond its current horizon should allow BASF to to look at innovation on multiple fronts to deliver further growth successfully in the years ahead.

SECTOR OVERVIEW

Traditional investment in the chemical sector has been in mature, industrialised countries but increased globalisation now means there is substantial activity taking place in Asian economies. Because commodities still have a significant share of the product portfolio, companies increasingly focus on innovation to stay ahead and provide a healthy financial return. Firms such as Dow, DuPont, BASF and Bayer are being increasingly challenged by changes in feedstock availability and price, rises in labour and utility costs, differential rates of economic growth and environmental pressures. While Europe still accounts for roughly two thirds of global chemical production and exports to China have been rising, this trend will probably reverse soon as the full Asian production capacity comes on line.

Instrumental in the changing structure of the global chemical industry has been the growing participation of regions such as the Middle East, South East Asia and specific developing markets such as Nigeria, Thailand and Venezuela. The driving factor for this is a desire for improving profitability by reducing production costs and an increasing demand for core product supply to be convenient to core future markets. Companies now choose location for a specific operation based on the levels of trade between adjacent countries and high competition for market share.

At the same time, the need for improved profitability from reducing production costs is being further driven by sustainability issues especially for those companies operating with Kyoto complying countries which have been facing the costs associated with decreasing their CO2 emissions. Lastly, the ever-mounting pressure for chemical companies to explore more sustainable feedstock alternatives and correspondingly enable a shift from the current dependency on petrochemicals has led to an increase in focus on white or industrial biotechnology solutions.

INNOVATION DRIVERS

The past five years has seen the introduction of the highest number of environmental regulations in the chemical industry's history and, with new legislation coming into effect, many companies across the sector are expected to develop alternatives for chemicals which are no longer permitted. Innovation and R&D is increasingly focused on developing alternative raw materials as a result of the rising oil prices which has spurred oil companies to invest in natural gas installations. Alternative materials are starting to include bio-based feed-stocks such as corn and many biotech companies are investigating more environmentally friendly processes to turn these into chemical platforms.

While this sector has not been traditionally overly concerned with consumer and market trends, increasingly several leading chemical companies have started to recognise the benefits to be gained from looking across the entire value chain and understand where and how to focus their attention. This is most apparent around such global trends as health and fitness where chemicals companies supplying the food and pharmaceutical sectors have risen to this challenge. With this realisation, like many others, the sector has been looking more externally for new ideas and technologies and some of the key players have adopted a more open system of innovation and are starting to see the results. However, in reality, in many companies most of the internal processes required to support driving a customer-centric perspective are not in place, so many of the potential benefits that lie there remain underexploited.

At a macro talent-supply level, a major challenge in Europe and the US has been the declining attractiveness of the chemical sector as a career choice: the number of graduates in chemistry has been in decline since the 1990s as the next generation are choosing alternative careers. In addition, there has also been a considerable 'brain drain' as experienced workers have left for other sectors. As China and India face the opposite trend, this is driving more and more companies to recognise that the future for innovation in terms of market opportunity and capability location lies outside Western markets.

ONES WE ARE WATCHING

Dow Chemical

US based Dow has a strong focus on science and technology in the areas of plastics, chemicals and agricultural products which continue to make it a formidable player in the chemical industry. The company's Innovation Team serves as the focal point and driver for Dow's innovation portfolio. With recent innovations in areas such as water purification, automotive, nutrition and construction Dow has successfully embraced the drive for sustainable solutions.

DSM

DSM has been pursuing a journey of transition over the years with notable success. As it first moved from coal and fertilizers on to performance chemicals, it saw the future challenges and has subsequently evolved into life-sciences and biomaterials. Now, through a combination of additional internal innovation, selective early-stage acquisition and venture activity, DSM is aiming to increase innovation output from €300m in 2007 to €1bn by 2010.

AUTOMOTIVE

Driving change in a highly competitive marketplace through continually enhancing the aspirational driving experience

BMW

Founded in 1917, the BMW Group is now one of the ten largest car manufacturers in the world and, with its BMW, MINI and Rolls-Royce brands, possesses three of the strongest premium brands in the car industry. The group also has a strong market position in the motorcycle sector and operates a successful financial services business. The company aims to generate profitable growth and above-average returns by focusing on the premium segments of the international automobile markets. With this in mind, a wide-ranging product and market offensive was initiated in 2001, which has resulted in the BMW Group expanding its product range considerably and strengthening its worldwide market position. The company's brand is extremely strong and is associated with high performance, engineering excellence and innovation. Indeed, the BMW brand is often cited as one of the 'best' in the world, and the company continues to launch a stream of innovative products as part of its battle with German peer Mercedes to be the world's largest luxury car maker.

Innovation Scorecard	
Innovation Culture	8
Strategic Innovation Focus	9
New Products	8
Managed Growth	9
Product Margin	8
Investment in Innovation	9
Innovation Brand Impact	9
Innovation Peer Review	9

BMW Key Data 2007	
Total Revenue	€56.0bn
Revenue Growth	14.3%
Profit Before Tax	€3.9bn
Worlwide Unit Sales	1.5m
Unit Sales Growth	9.2%
Profit / Unit	€2,580
Total Employees	107,539
Revenue / Employee	€521k

BMW's focus on engineering excellence allied to leading-edge design continues to drive successful, profitable expansion. In 2007 BMW sales increased by 8%, Mini by 18% and Rolls-Royce by 26% with, for the first time ever, over 1000 of the super luxury cars being produced in one year. To further this growth, a host of new models is being launched, including the Mini Clubman and the new sport utility vehicle, the BMW X6 - the world's first SUV coupe. While the Clubman reinvents views on vehicle access, the X6 is an excellent example of BMW innovation at work. It combines the safety and convenience of a four-wheel-drive with the on-road performance of a sports car and is designed to appeal to the driver who enjoys a commanding driving position, but also savours the characteristics of a sports car. With its stretched coupé silhouette and pronounced performance design, underpinned by hybrid engine options, as previously achieved with the X5 and the X3 in allied markets, the X6 is the latest instance of BMW changing perceptions of what a car should provide – for its passengers and its driver alike. At its heart, it restates an aspiration for driving that is both exclusive and yet also available to the mass market.

BMW has also been at the forefront of introducing new IT options to enhance the driving experience. Starting with the iDrive first introduced in the 7-series, BMW ConnectedDrive is now available across most models and is adding greater functionaility. After being one of the first to offer the capability for MP3 connectivity and incorporate RSS feeds including weather information, in 2007 BMW teamed up with Google to offer a PC driven route planning service. Of course this level of innovation does not come cheaply and a key challenge going forward will be to keep research and development costs under control. During the last five years, BMW's average annual R&D investment has been around €2,300 per car, compared with €1,700 spent by arch-rival Mercedes. Alongside the examples above, much of the money has gone into the car maker's Efficient Dynamics programme aimed at making engines more efficient, improving aerodynamics, reducing weight and capturing energy during braking. As the numbers clearly show, BMW is a mass market player but one that successfully uses focused innovation to build and maintain the aspirational driving experience for many. Through a constant stream of consumer informed innovations, the company has moved ahead of its peers and future sustained and profitable growth is widely predicted.

SECTOR OVERVIEW

Following a century of non-stop growth there is large scale over-capacity in much of the motor industry. Widespread incentives and longer-term warranty programmes have been applying pressure, and financial strains are increasing due to closer partnerships between vehicle manufacturers and suppliers, higher commodity prices and sub-sector consolidation. With little recent significant M&A activity, growth in vehicle production has largely been organic . As the US big three struggle to be profitable, other established players such as Honda and Toyota have focused on steady growth by developing, producing and distributing new models with increasing efficiency. Unlike their counterparts, they have avoided price reductions, choosing instead to introduce an evolving range of high-quality vehicles and new design-led sub-brands such as Lexus and Infiniti. This may seem like incremental rather than radical change, but underneath continuous technological and consumer-led innovation is taking place.

The main development of note at the moment is in emerging markets where both China and India are fast moving from follower to leader status. Having been traditional markets for established manufacturers to offload old designs, the tables are turning as Chinese and Indian brands emerge into the world market with an ambition to pioneer new innovation. For example, it is forecast that China will produce 130 million vehicles in the next 20 years – by contrast Ford has only produced 90 million in the past century. Such volumes hitting already crowded streets demand advanced innovation to simultaneously satisy the growing middle class whilst avoiding gridlock. Although intelligent vehicles and hydrogen fuel cells have been pushed by US and European manufacturers over the past few years, many commentators now see that it will be in China and India that such developments occur first on a mass scale and, in so doing, allow domestic manufacturers to leapfrog the established multinationals.

INNOVATION DRIVERS

The most visible driver for innovation of 2007 was the need for greater efficiency and the move to a greener form of automobile. All the main auto shows were dominated by the large automakers showing their plans for cars with alternative energy sources or for improved traditional systems. Many analysts see that the car makers are pushing to develop environmentally friendly cars because they fear that their link with gas-guzzling vehicles is not helping their image. Whilst this is true, the main reason for this innovation is the statutory requirements within which automakers must work with regards to CO2 emissions. With the EU imposing a 130g/km average limit for new passenger cars by 2012 and California looking to move to a 30% reduction in emissions by 2016, regulation is now just as much a driver of change as is consumer demand.

In terms of design the impact of these measures will not only be felt in the engine design area, but also inside the cabin where the physical space of the car is being reconfigured to account for different shapes of batteries and fuel cells. Alongside developments in hybrid engines led by Honda and Toyota, most of the major automakers are experimenting with new hydrogen fuel-cell-based platforms. At the same time, firms such as Renault and Ford-owned Volvo have been introducing a stream of design innovations to improve passenger and now pedestrian safety.

In addition, the full exploitation of multiple-variant opportunities for each platform, which meet niche consumer requirements with a host of space, performance and comfort improvements, has also been a major recent area of innovation activity to support competitive positioning. Finally there are also shifting trends in fuel-type usage in the key geographies that will drive innovation in the whole supply chain: the US moving towards more hybrid use, Europe going more for bio diesel and South America embracing the bio ethanol option. Although many are now questioning the sustainability of the accelerating move to first generation bio fuels, the long-term trend to a broader mix of fuels is clear. As second generation bio fuels, which do not clash in the food vs. fuel debate come on line, they will be joined by more advanced electric power options that take advantage of new battery technologies.

ONES WE ARE WATCHING

GM

Although still suffering the impact of legacy financial obligations, there are several signs of GM's innovation-led recovery. Alongside sustained growth of the Saturn brand in the US, new models in 2007 included the Chevrolet Volt concept car - the first mass-market prototype designed to operate purely using an electric motor. In addition, as it focuses on delivering a stream of mass-market incremental innovations in response to new insights on use of the family vehicle, the Meriva Concept has been built on the theme of interior flexibility. With doubled sales in Russia, more than 1m vehicles sold in China and 500,000 in Brazil, GM's global brands are starting to fight back against Toyota.

Tata

As well as being the new owner of the premium Land Rover and Jaguar brands, India's Tata Motors recently unveiled the world's cheapest car, the Tata Nano. Rather than offering low-performance versions of existing models as other have done, this product fundamentally rethinks what a mass-market car should be and delivers it as a complete, accessible, low cost, high volume package. Providing a 'car for the people', the Tata Nano uses extreme simplicity at the heart of its strategic intent, while at a global level, as a potential breakthrough platform innovation, the repercussions as versions 2 and 3 come into play could be huge.

Bringing resource-efficient, high-tech mass production to a sector traditionally focused on batch manufacture

BOEING

You can't but be impressed by the scale of the Boeing organisation: it is the world's leading aerospace company and the largest manufacturer of commercial airliners and military aircraft. Boeing also designs and manufactures electronic and defence systems, missiles, satellites, launch vehicles and communication systems. Furthermore, as a major service provider to NASA, the company also operates the Space Shuttle as well as the International Space Station. With customers in more than 90 countries and revenues of over $66bn, Boeing has aerospace covered. Although it faces strong competition in some areas, as a whole, no other organisation comes close to Boeing's breadth of activity, but that does not make the company complacent. Faced with challengers from within and outside the US, it has responded with a focus on innovation-driven growth across its portfolio and is looking good for the future.

Innovation Scorecard

Innovation Scorecard	
Innovation Culture	7
Strategic Innovation Focus	7
New Products	8
Managed Growth	7
Product Margin	6
Investment in Innovation	7
Innovation Brand Impact	8
Innovation Peer Review	8

Boeing Key Data 2007

Boeing Key Data 2007	
Total Revenues	$66.4bn
Net Earnings	$4.1bn
Operating Margin	8.8%
Airplane Sales	$33.4bn
Airplane Deliveries	441
Gross Airplane Orders	1423
#US Patents	436
Revenue / Employee	$443k

Within Boeing Commercial Aircraft, the big news story of recent years has been the 787 Dreamliner which, with another 369 orders in 2007 and 800 since 2004, is the fastest-selling new airplane in history. Although not yet launched and going through some of the usual supply chain challenges for innovation in the sector, this new airplane is not just another me-too product that is slightly bigger, smaller or faster than what has gone before. In a sector where batch production is more common than mass production and individual hand finishing is the norm, the 787 has been designed to take advantage of the latest composite materials to both enable the adoption of more mass-production techniques and also save weight – an increasingly important issue. Alongside the 787, Boeing has just passed key milestones including the seven-thousandth order for the 737 and the one-thousandth order for each of the 767 and 777.

On the defence side of the business, Boeing is winning the majority of new contracts. Although military orders for aircraft are generally declining, new programmes that have been developed to provide support activities to customers are going down well. These include new border security systems, support for NASA's Ares I programme to build the next generation of tracking satellites and new maintenance programmes.

Underpinning growth in both parts of the business has been the pivotal role of a new CEO and several major initiatives focused on gaining extra efficiency and effectiveness from the organisation. These have included Lean+ which is accelerating the application of lean principles; Development Process Excellence which is increasing speed and yield of research and development activities; Global Sourcing which is leveraging the purchasing power of the entire company across the worldwide supply chain; and, most recently, for the first time a true company-wide research and development strategy. In this, the most technologically sophisticated of companies delivering high impact product innovation, internally focused process innovation is arguably the primary driver of overall profitability.

SECTOR OVERVIEW

In the commercial airliner market, Boeing and Airbus have been head to head for several years now, each trying to outperform the other in terms of both new orders and deliveries. Airbus has been pushing the A380 'super-jumbo' as the ultimate large-scale airliner capable of moving hundreds of people at a time between major hubs, while Boeing have gone for the 787 Dreamliner which is designed for arguably more flexible and more economic point to point routes. Both have won orders, both have experienced delays and both are keen to promote their individual benefits. Airbus was first to market with fly-by-wire electronics that revolutionised the cockpit while Boeing has been pushing new materials as a key differentiator. However, while these two have been battling with each other both openly and behind the scenes in political circles, other upstarts have been making a strong push in attractive segments. Most notably Bombardier and Embraer have been successfully launching new products into the fast growing regional-jet markets.

In the military world, flattening defence budgets, the escalating costs of Iraq and Afghanistan and shifting future priorities for the world's major powers are all having an impact on the sector. Although the largest military order in history - that for the $200bn Joint Strike Fighter contract is keeping Lockheed and its main suppliers busy, large-scale, future manned fighter development is open to question. With the use of sophisticated unmanned drones such as the Predator increasingly being used in many conflict areas, the appetite for risking pilots' lives is on the decline. In addition, on the back of the huge success of the Mars Rover robotic vehicles, even NASA is favouring unmanned space flight for much of the near-term future exploration projects. Although the International Space Station and potential moon bases keep astronauts in the public eye, for the majority of applications robots are cheaper and more reliable.

Across both segments of the sector, there are also a number of principal suppliers which have increasing influence. These include the likes of Smiths Industries and Honeywell in the avionics arena and GE and Rolls-Royce in the design and manufacture of engines. Both of these companies have been highly successful in creating significant after-sales support services such as TotalCare which now accounts for over half of Rolls-Royce's revenues and most of its profits. Even in this sector, innovation in services is having as much impact as product and technology innovation.

INNOVATION DRIVERS

Energy consumption is the top preoccupation for commercial and military plane manufacturers alike. With rocketing oil prices and public concerns about carbon footprints, even the US Air Force has started to use energy efficiency as a primary criterion in its future purchasing strategy. As well as driving intensive innovation for the likes of GE and Rolls-Royce in terms of the fuel efficiency of their engines, this has also focused the minds of airplane manufacturers and airlines alike on the weight of the planes, the seats and the controls. More that any other issue right now, doing whatever it takes to reduce fuel consumption in an environmentally efficient manner is the critical concern across the sector.

Alongside this issue, the other two major drivers of innovation in the sector are NASA and the Pentagon. With the US still leading in space exploration and satellite production, and US military spend outstripping the rest of the world put together, given the scale of the contracts available, what NASA and the Pentagon want to achieve is top of the priority list for many aerospace companies. Not just in the military arena but also in the commercial airlines market, technology development in areas such as virtual engineering, simulation and modelling, intelligent systems, advanced smart materials, safer airframes and more reliable guidance systems are all being driven by specifications defined by these two bodies.

Correspondingly it is no surprise that, from a competitive perspective, there is a significant role for the EU in supporting an independent aerospace sector. While Brazil, India and China all have varied commercial and military aerospace activities, in terms of scale only the likes of Rolls-Royce, BAE Systems and, most significantly, EADS have the ability to compete with the major US based firms. The European Space Agency and the governments of the UK, France and Germany in particular are all keen to have innovation active within the EU and so support a number of major programmes. That said, for the European aerospace companies, the US is still the primary market, the one that sets the agenda and so, politics not withstanding, is pivotal in innovation policy.

ONES WE ARE WATCHING

Embraer

Embraer, the Brazilian manufacturer of regional jets, is going from strength to strength and has now sold over 1000 of its ERJ product. With net sales of over $5bn in 2007 and net income of just under $0.5bn, its 170/190 ranges of commercial airliners have been a huge success with new orders from airlines including KLM, Lufthansa, Virgin Nigeria and TACA. Moreover, riding the fractional ownership wave created by the likes of Net-Jets in the rapidly expanding executive jet market, Embraer's Phenom family now has over 700 orders. Add to this the recent creation of an after-sales services business, and Embraer is now a major player on the world aviation stage.

EADS

Airbus manufacturer EADS is again in the midst of politically driven reorganisation but it has some excellent new products. Unlike the 787, the A380 is flying, albeit a bit later than promised, and orders for this mammoth product are being fulfilled. At this same time the A350 is in development and, in the military arena, EADS's probable success in winning the second-biggest military procurement contract of all time for aerial-refuelling tankers is causing ructions in Washington. Yes, EADS is allegedly subsidised by the French and German governments but the Pentagon usually gives Boeing a helping hand. EADS is resurgent and, with a proactive policy towards China, tipped for growth.

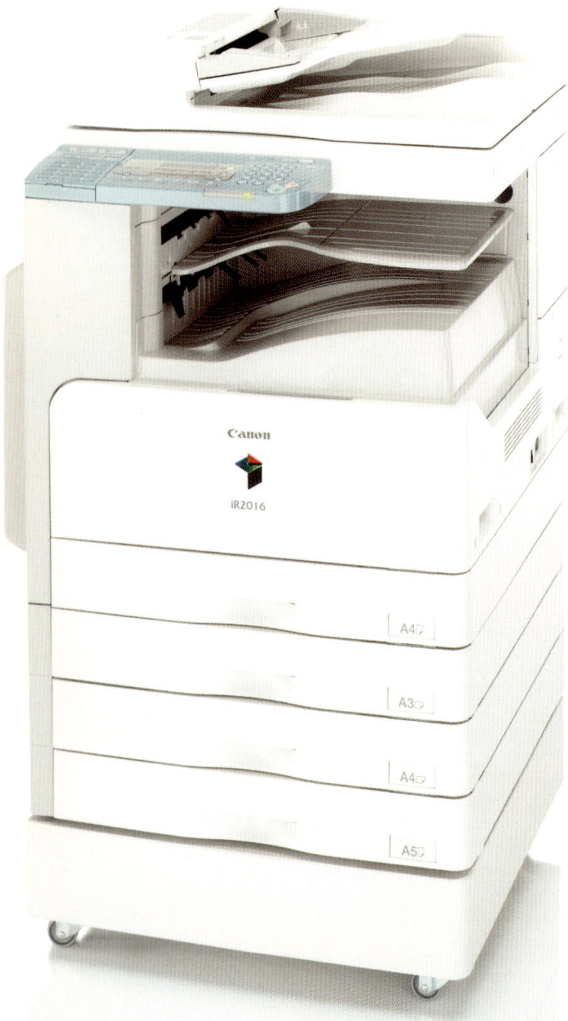

OFFICE EQUIPMENT

Exploiting core technology competencies to chosen market opportunities to deliver an integrated ecosystem

CANON

Canon, known by many for its cameras, is also the world's leading supplier of copying machines, laser printers and paper management systems. With continued growth in 2007, it is also one of the few multinationals that has maintained a long-term corporate culture focus on a balance of success and responsibility. In living to fulfil its corporate philosophy of 'kyosei', Canon strives to contribute to the prosperity of the world and the happiness of humanity, leading to continuing growth while bringing the world closer to achieving total harmony. A fundamental belief is that true global companies must foster good relations, not only with their customers and the communities in which they operate, but also with nations and the environment. They must also bear the responsibility for the impact of their activities on society. In a sector partly focused on the ideal of bringing the 'paperless office' closer to reality, this is an interesting position. Specific targets have been set for the ratio of net sales to lifecycle CO2 emissions from the company's activities by 2010. To achieve this objective, Canon is currently busy adapting its technological development processes and business operations.

Canon

Innovation Scorecard	
Innovation Culture	8
Strategic Innovation Focus	8
New Products	8
Managed Growth	7
Product Margin	8
Investment in Innovation	9
Innovation Brand Impact	8
Innovation Peer Review	8

Canon Key Data 2007	
Net Sales	$39.3bn
Sales Growth	7.8%
Net Income	$4.3bn
Net Income Growth	7.2%
R&D Investment	$3.22bn
R&D as % of sales	8.2%
# US Patents	2037
R&D / Patent	$1.6m

The company has focused much of its development effort on areas it believes can support its corporate philosophy. These areas include 'prototypeless design' where development costs and lead-times have been considerably reduced by the replacement of multiple physical prototypes with 3D CAD systems, simulation, measurement, and analysis technologies. In addition, the company has invested heavily in the development of its 'Five Imaging Engines' - namely image capture, electrophotography, inkjet printing, photolithography and display technology to support the Canon product businesses. Each engine is strengthened by a fusion of various technologies and has become the source of the company's competitive advantage. At the same time, Canon has continued to pioneer its four core-competency areas of precision mechanics, fine optics, microelectronics and electronic imaging, resulting in the development of innovations such as CMOS Sensors, the electronic eyes of digital cameras and Color Management, the unified system used for all Canon products that have embedded colour imaging. Just over 8% of the company's sales is ploughed back into new technology research, which is undertaken by a global R&D network spread over seven locations worldwide. One outcome is 14 consecutive years as one of the top 3 recipients of US patents.

Supporting this R&D is a considerable market effort to distinguish Canon from the competition. Global relationships have been built on a Java-based open-application platform that enables the addition of functions to complex office machines including WSSS - a web-based system that provides online 24-hour, year-round customer service and support for Canon products. In addition to ensure continual evolution of products around customer needs, direct consumer input to innovation is monitored by a system that keeps Canon abreast of product and service quality information through centrally managing comments received from customers. The data compiled is shared worldwide and use in product development and improved user guides. With this combination of radical and incremental innovation being driven equally by strategic intent and cultural philosophy, Canon is widely regarded as a leader in sustainable growth and one which many are seeking to emulate. 2008 is the start of the next phase in Canon's evolution as it aims to achieve the overwhelming number 1 position in all core current businesses while simultaneously expanding the company's activities into new domains. To sustain this growth, Canon is expanding on its Five Imaging Engines and creating future business based on new engines by boosting basic research and key-component technology development.

SECTOR OVERVIEW

The office equipment sector is evolving rapidly and sits at the intersection between consumer and business electronics and the IT industry. The sector is populated by such companies as Konica Minolta, Océ, Brother and Canon. With the market in developed countries near saturation many companies have been turning to new opportunities. The expanded European Union has been a particularly important growth area for printers and copiers as well as multifunctional peripherals. Meanwhile in more mature markets demand has been stimulated by the advent of high-end multifunctional digital devices, developed specifically for intranet / internet office environments.

These markets are also seeing an increased demand for high-resolution colour printing. The continued trend towards more home working and the mobile office means there is a constant emphasis on the need for portable, lightweight peripherals as personal computers become smaller and converge with the PDA. As with other sectors, as the market and the sophistication of the products increases, their physical size decreases. Finally, another, less talked about, influence on the office environment is the strengthening and implementation of country-specific disability acts that are targeting reduced employment discrimination against those with disabilities.

INNOVATION DRIVERS

The strongest innovation driver in this sector has been and will continue to be digitisation. The example of the network printer that can efficiently and effectively operate as a scanner, copier, fax machine and even be used for emailing is emblematic. Originally print-control attachments were developed to allow copiers to become multifunctional; now printers can be converted to provide PCs with all the required functionality. Inkjet printers are being further eclipsed by the demand for better quality and faster colour print, with laser printing now costing significantly less than on the average inkjet printer. All of the recent developments have created a need for advanced management software for office automation equipment alongside the ability to remotely administer the equipment and, where necessary, effect repairs and modifications. Another driver has been the need to extend the paperless office through document storage and retrieval solutions at the low-end of the market, specifically targeting the small office and home working with appropriately sized equipment.

ONES WE ARE WATCHING

Brother

Established in 1934 to manufacture and distribute sewing machines, Japan's Brother Industries' product mix now also includes machine tools and information and communications equipment such as printers, facsimiles and multi-function centers. Focused very much at the value-end of the market, Brother has expanded its business operations overseas, with 19 production facilities and 41 sales companies in 39 countries and regions, and in 2007 this delivered 562bn Yen. The company is currently in the second of three phases in a 10 year programme where driving growth is the priority, and it is making large capital expenditures and substantial investments in R&D programmes.

Xerox

Having stalled in recent years, Xerox is becoming an innovation powerhouse again, producing new technologies that can read, understand, route, and protect documents. It provides the industry's broadest portfolio of offerings. Digital systems include colour and black-and-white printing and publishing systems, digital presses and 'book factories', multifunction devices, laser and solid-ink network printers and copiers, as well as associated software, support and supplies such as toner and paper. With 2007 revenues of $17.2bn, Xerox commited 5% of this to R&D with a focus on six S's: making systems simpler, speedier, smaller, smarter, more secure, and socially responsible.

Google is using perpetual beta to drive disruption in multiple areas while consistently building core advertising revenues

GOOGLE

Widely seen as the world's largest search engine, as its core offer Google provides the world with an easy-to-use, free service that continually returns relevant results in a fraction of a second. However, the core of the highly successful, highly profitable and highly valued business that is Google is actually its role in refining the advertising industry. Amongst an ever increasing diaspora of multiple new product and service introductions, at heart, Google's primary focus is to use most of these to drive increased traffic to its ever expanding online, mobile, and soon to be offline, advertising revenue streams. This is clearly paying off as 2007 annual revenues increased to over $16bn, net income increased to over $4bn, Google's share price continues to beat Wall Street's expectations and, much to the irritation of Microsoft and Yahoo, through AdSense and AdWords the company commands an impressive 32% market share in online advertising.

Google™

Innovation Scorecard	
Innovation Culture	9
Strategic Innovation Focus	9
New Products	9
Managed Growth	8
Product Margin	8
Investment in Innovation	9
Innovation Brand Impact	10
Innovation Peer Review	10

Google Key Data 2007	
Net Revenues	$16.6bn
Net Income	$4.2bn
Operating Profit Margin	34.2%
Brand Value	$17.8bn
R&D Investment	$2.1bn
R&D Intensity	12.8%
# US Patents	37
Revenue / Employee	$987k

Testament to the success of Google is the common use of the brand name as a verb and, in 2007, the company continued its mission to make information universally accessible through an increasing range of services. Google is digitizing books at the rate of 3000 a day and has been launching a diverse range of services from patent search to local discount coupons, searching for government information, pay-per-action ads and simply providing Google News on mobile devices.

Alongside the growth of disruptive services such as Google Finance, Google Scholar and Google Desktop, the increasingly global adoption of the likes of Google Maps, Google Earth and Blogger has made such services the defacto standards. Much to the concern of the content-owning companies being challenged by Google's increasing reach, has been the company's exceptionally successful use of perpetual beta. Unlike many established players who seek to get a new product or service as close to perfect before launch, Google is willing to launch beta versions of its innovations into the market earlier and let users influence and drive configuration. So attractive is this to the company and its customers that some products are in continuous development, are never finished and so are perpetually in beta. Google can do this but, right now, many of the companies whose markets it is disrupting can't.

2007 developments of note included Google Checkout, which makes it easier to buy products advertised on Google, and the release of OpenSocial to tap into the social networks and introduce more targeted advertising. This is proving successful and has commitments from community sites such as Linked-In and Ning as well as companies including Oracle and Salesforce.com. In the wake of the acquisition of YouTube, which is changing viewing habits around the world, high-profile purchases included the DoubleClick deal at $3.1bn. This is seen as much a defensive move to prevent others becoming a threat as it is a way for Google to grow its revenue by applying targeting capabilities to display ads. However with the deepest pockets in the marketplace, this scale of purchase should not be a surprise. Google is also a successful practitioner of smart M&A where notable acquisitions in 2007 included Postini, GrandCentral Communications and Endoxon.

Going forward, Google is targeting the mobile space. Strategic partnerships with telecom operators, an increasingly high-impact cooperation with Nokia, bidding in an auction for wireless airwaves and plans to develop its own operating system are all indications of some of Google's intentions: mobile versions of YouTube and Google Maps are fast becoming the standard in this growing marketplace.

SECTOR OVERVIEW

Today's media landscape is partly characterised by big multinational companies with wide-ranging activities covering magazines, television, radio, film and books as a result of a tradition in diversification and acquisition. On the other hand there are organisations such as Google, Apple and Yahoo who have been seen to innovate across sectors and have an impact on media in a way not foreseen by the incumbents. With the introduction of Web 2.0 the focus has again shifted. This time it's networks and communities that are catching everybody's attention as demonstrated by NewsCorp's acquisition of MySpace, Yahoo's interest in Facebook and Google's acquisition of YouTube.

Online video is gaining traction, Time Warner successfully broadcasted Live8, YouTube was streaming 100m videos each day and Disney sold half a million films through Apple's iTunes within the first 2 months of their distribution deal. Apple announced plans to shake up the online movie rental business by offering movies from most major studios through iTunes. At the same time Hollywood was able to continue its recovery from a decline in movie attendance in 2005. Box office receipts were up by 7.2% compared to 2006 with hit movies such as Harry Potter and third instalments of Pirates, Spiderman and Shrek. But while DVDs generate over 50% of the revenues from most movies, sales growth remained flat and the expected pickup from high definition DVDs was slowed due to technical delays and a format war between consortia led by Sony and Toshiba that was finally resolved with Blu-Ray as the victor in January 2008.

While it is expected that consumers will spend more on media and entertainment over the years to come, they will also be more demanding in terms of price, convenience and speed.

INNOVATION DRIVERS

The media sector was one of the most profitable ones during the 20th, century when today's incumbents were able to reach mass audiences through broadcasting and publishing. Today the sector is characterised by more and more fragmented audiences who refuse to be segmented based on traditional features and are increasingly more demanding in their media consumption. This calls for a different approach, one which seems to still elude yesterday's media giants but is a natural fit for new media revolutionaries. While people are spending more time using the internet as their main route to information and entertainment the 'old' distribution channels are still quite profitable and this poses a barrier to change. However, there are signs that media organisations are picking up on trends such as networks and communities where new business models emerge. MySpace is having an impact on the music value chain and Wikipedia is redefining the way encyclopedias are produced. The classic distinction between professionally produced media and content generated and distributed buy amateurs seems to be gone forever. In fact the reverse is true in some cases where for instance blogging is now picked up by respectable newspapers such as the *Washington Post* and the *Guardian* in an effort to capture audiences and build relationships that otherwise would have been lost. The challenge remains for organisations to either use strategic innovation to develop new profitable business models and build innovation capability or to identify and invest in new opportunities as they emerge at an early stage.

ONES WE ARE WATCHING

News Corporation

As it continues to extend its reach in all of its TV, print and online markets, News Corp remains a strong contender for the future leadership with strong financial growth performance over the last few years. As well as buying Dow Jones & Co, in 2007 the company delivered growth across the board, including online where MySpace turned a profit for the first time. Given its leadership position in direct satellite broadcasting and its ability to syndicate film, text and TV content across its many newspapers, book and magazine publishing and online channels, News Corp is a major player in innovating the future of media.

Disney

With Pixar now firmly in the family, $35.5bn revenue Disney is growing well on all fronts. Across domestic, satellite, cable, mobile and internet, its media broadcast businesses are extending reach: the increasing range of parks, resorts and vacations around the world are must-do for many kids, studio entertainment across both animated and live-action areas continues to deliver hits like *Ratatouille*, *Enchanted* and *Pirates of the Caribbean 3*, while underpinning it all, Disney Stores in key markets are generating additional revenue streams. Knitting all this together has been a challenge but cross-channel innovation at Disney is again in full swing.

Exploiting the road less travelled of decentralised local relationships in an era of centralised commoditisation

HANDELSBANKEN

Handelsbanken is a traditional Swedish Bank offering corporate transactions, investment banking and trading as well as consumer banking, including life insurance. Its stated aim is to provide a first-class banking service to clients who value what might be described as 'old fashioned' banking. In global terms the company is a relatively small bank but in terms of the key measures that count it performs remarkably well. It has approx 11,000 staff operating in 457 branches in Sweden, 162 in other Nordic countries and Great Britain as well as units in 17 other countries. Handelsbanken has total assets of SEK1,859bn and in 2007 made an operating profit of SEK19.4bn.

In a world of increasing centralised commodification of core products, Handelsbanken's business model is simple and highly effective: 'the branch is the bank' and the corporate slogan clearly defines the philosophy 'banking is global – business is local'. Handelsbanken has made personal service and decentralised decision making the cornerstone of its banking strategy. Each branch is an independent profit centre, not one channel among many.

Handelsbanken

Innovation Scorecard	
Innovation Culture	7
Strategic Innovation Focus	8
New Products	8
Managed Growth	9
Product Margin	9
Investment in Innovation	8
Innovation Brand Impact	7
Innovation Peer Review	8

Handlesbanken Key Data 2007	
Total Income	*SEK32.6bn*
Income Growth	*10%*
Operating Profit	*SEK19.4bn*
Profit Growth	*13%*
Return on Equity	*20.9%*
Total Assets	*SEK1,859 bn*
Cost / Income Ratio	*40.5%*
Income / Employee	*SEK3.1m*

The whole bank is essentially managed 'bottom up' with a focus on monthly P&L and the benchmarking of key ratios. The local manager is responsible for developing a business plan and makes all key decisions including staffing, competence profiles, customer selection, product mix and pricing. And importantly the responsibility for marketing is also held by the branch - it selects customers, is close to their needs and hence the focus is on customer profitability - not product profitability. This has resulted in 98% of credit decisions being made at branch level and a subsequential lower bad credit rating than many of its peers. This local and simple philosophy has carried over into its approach to internet banking. Each branch has its own website and email address. With high penetration rates of 40% in consumer space and over 70% in the corporate market, a tremendous reduction of workload and increase in overall efficiency has been consequently achieved.

To highlight how different Handelsbanken is, it is useful to compare it with the 'standard' approach taken by the large international retailers: Handelsbanken is decentralised and flat whereas its competitors tend to be centralised and hierarchical. It also focuses on returns and relationships, not volumes and products.

But what has been the effect on overall performance? Return on shareholders' equity during the period 1973-2005 exceeded the weighted average of comparable listed banks for each year. Handelsbanken is also able to manage the seemingly incompatible goals of a better service with a lower cost and having, for example, the most satisfied customers in its markets. In terms of cost-effectiveness it was shown to be the best performing bank in a recent study of European universal banks: with a cost / income ratio of 40.5% Handelsbanken is certainly efficient.

Another of Handelsbanken's unique features is its focus on low credit losses, achieved through a unique credit granting process requiring a commitment in terms of both competence and processing times: customers are chosen selectively, analysis is done close to customers and money cannot be borrowed unless approved by the branch; this leads to high quality borrowers and long-term customer relationships. Innovation at Handelsbanken really is a case of 'Back to the Future' through using appropriate technology to support trusted and highly efficient relationship-based banking. As the full extent of the credit crunch becomes apparent it may be that others will follow.

SECTOR OVERVIEW

It is commonly said that retail banks are poor at innovation but this is not entirely fair. Innovation does occur but much of it is hidden, focusing more on new ways of doing business, rather than the development of new technology and products. Innovation in many banks is internal to the organisation and largely invisible to the customer. That said, many retail banks are, in the main, well behind the innovation game.

Most examples of retail banking innovation occupy the blurred region between process and product. Take for example the development of payment methods. The way consumers transact purchases is gradually changing; from the slow demise of cheques, to the wide availability of credit cards, debit cards and chip-and-pin and the new ways of making 'contactless' payment. Even that old stalwart, cash, is becoming marginalised. As electronic cash finally starts to make inroads, contactless cards are now the norm in some cities. Hong Kong's Octopus card, a contactless stored-value smart card originally designed to enable payments on the city's public transport system, is 10 years old and is now used at a wide range of retail outlets. Similarly over 5 million people regularly use Transport for London's Oyster card, with over 10 million issued since its launch in 2003. In Europe contactless payment technology is beginning to be used as a way of making regular payments for small retail transactions; for example, customers using Visa's 'payWave' system simply hold their card up to a secure reader to make their payment. In the UK, Barclays Bank's OnePulse card, launched in autumn 2007, combines a credit card, cashless payment and Oyster travelcard in one product.

Alongside smart card-based transactions, payment by mobile phone using similar near-field communication technology is already available in Japan and Korea and, at the other end of the spectrum, Vodafone's M-Pesa money transfer service is having a major impact in Africa. Given the pervasive nature and rapid adoption of mobile technology, widespread mobile payments are likely to scale up rapidly in other markets in the near future.

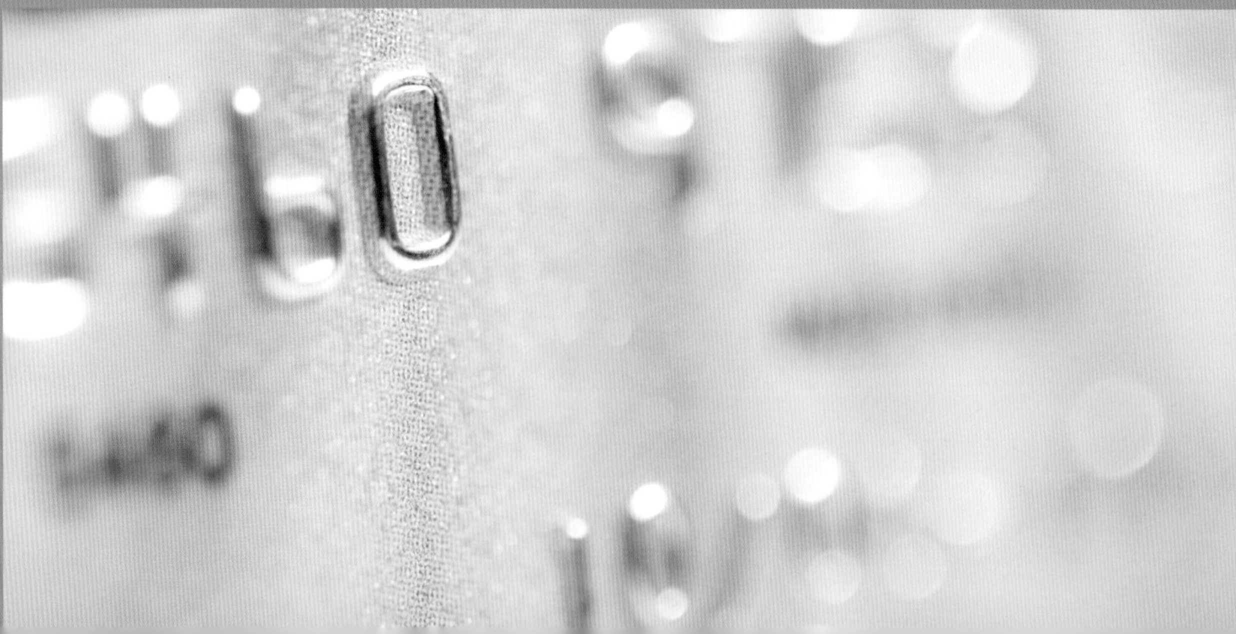

INNOVATION DRIVERS

The retail banks' organisational model is undergoing continual change. The last two decades have seen ongoing evolution from a simple product focus to segmented distribution by customer type through multiple channels, central design of products and determination of credit, and efficient operations serving a range of frontline businesses. Globalisation and internationalisation continue to reduce the number of separate banks. Middle-ranking banks, in particular, lacking economies of scale, are likely to be squeezed by the large multinationals on one side and by small niche providers on the other.

As with several other sectors, regulation is a major factor in driving both innovation priorities and market trends. Varied national bodies and international agreements impose a range of onerous requirements, from treating customers fairly to effective business controls and prudent balance sheet management. The regulators are themselves scrutinised by consumer bodies, the government and by an often hostile media. In the major banking centres of the world, regulators are pursuing parallel agendas aimed at improving transparency in financial reporting, preventing money laundering and improving consumer protection.

Alongside such market change, the growth in the dependence of banks on IT systems is constant and relentless. Compliance and core processing have taken priority, since without these the bank cannot survive. However, going forward the challenge will increasingly be to maintain and develop, alongside these core systems, applications which can add real commercial benefit: systems which provide better information and are faster to allow smarter business decisions to be taken, systems which facilitate interaction with the customer and the creation of products for which there is unmet need.

Lastly, customers, as in many other consumer sectors, are becoming increasingly discriminating. The internet gives them access to vast amounts of comparative information, helpfully consolidated for them by price comparison websites. Customer attitudes are fragmenting, with individual references generating scores of niche interest markets. The ease of price comparison and simple transfer processes between suppliers, reinforced by intense competition, means both that brand loyalty is declining rapidly and that inertia is becoming a much less significant factor in customer retention.

ONES WE ARE WATCHING

Barclays

Barclays is gaining first mover advantage in UK converged card services. The OnePulse card, launched in 2007, combines a credit card, cashless payment and travelcard in one product. Launched into the London market, such multi-use cards are destined for mass take-up. Barclaycard OnePulse has three integrated functions - it is a standard 'Oyster' contactless travelcard for cheap travel around London, it is a standard credit card, and it is a cashless payment card whose 'OneTouch' payments allow for small purchases at over 1000 participating retailers.

Zopa

A radical entrant to the market, Zopa offers a new business model based on peer-to-peer 'social lending', where an online marketplace allows lenders and borrowers to be matched, bypassing banks completely. As yet, the sums involved are small but Zopa's model is already being copied in the US, Germany and the Netherlands. Whether social lending will pose a serious challenge to conventional retail banking in the near term is unclear but with 'no banks in the middle, no huge overheads, no unethical investments Zopa could permanently change the way people save and invest their money'.

Focused internal process innovation delivering the best fast-fashion product mix for customers in all locations

H&M

H&M has been at the forefront of affordable chic for some time now. From its Swedish base, Hennes and Mauritz has grown into a major multinational clothes and cosmetics retailer. Just the place for fashionistas, it now has over 1500 outlets in 28 countries and a turnover in 2007 of over SEK92bn. H&M was one of the first to challenge the retail clothing sector by delivering fast fashion at low prices. The company offers different concepts for women, men, teenagers and children and includes everything from modern basics to high fashion. The collections are supplemented by matching accessories, nightwear, underwear and cosmetics. However H&M does not have factories of its own but has a multitude of designers and buyers and works with around 700 independent suppliers to produce clothing collections for almost everyone, all at affordable prices. H&M lives by its business concept, 'fashion and quality at the best price'.

H&M

Innovation Scorecard	
Innovation Culture	8
Strategic Innovation Focus	8
New Products	8
Managed Growth	9
Product Margin	9
Investment in Innovation	7
Innovation Brand Impact	8
Innovation Peer Review	9

H&M Key Data 2007	
Total Revenue	SEK92.1bn
Revenue Growth	15%
Operating Profit	SEK18.4bn
Operating Margin	23.5%
# Stores	1522
# New Stores	177
Revenue / Store	SEK60.5m
Net Revenue / Employee	SEK1.7m

Expansion at H&M is the name of the game with 168 new stores being opened last year. The availability of attractive business locations is the major deciding factor but most recently there has been specific focus on the US, Spain, Germany, France and Canada. That said, a weather eye is also being kept on emerging markets such as Eastern Europe, China and Japan. The company's expansion has been entirely self-financed and with sales over the past 5 years increasing by 72% and earnings per share by 183%, the company can certainly afford it.

H&M is not solely reliant on new fashion store openings however: it also aims to expand its product range organically. Alongside launching H&M home, extending its ladies footwear range and growing its online shopping, it has recently launched a new chain called Collection of Style, or COS, which is an upscale version of its current H&M stores which merges high fashion with extremely competitive pricing. H&M mixes cool organised efficiency with zany fashion pizzazz. Or, to put it more practically, they have combined good basics with the hot-off-the-catwalk designs which keep people talking – all at a low prices. Store displays are designed to inspire but the essence of H&M is that it allows customers to find and experiment with their own style, irrespective of occasion and on a limited budget. Turnover is fast and stock is changed on a daily basis.

Although global in reach, H&M also recognises the power of local. Whereas other firms provide a ubiquitous retail experience with the same product range in San Francisco, Paris and Tokyo, H&M thrives on tailoring the ever-changing product mix to local tastes and buying preferences. This is attributed to the company's fast feedback from each store into a central IT and logistics operation that pushes the right products out from its Hamburg distribution hub. In a world where product innovation is the apparent differentiator, H&M also gains from expertise in process innovation.

Collaboration has also worked well for H&M. Following on from the breakthrough Karl Largerfield range in 2004, subsequent celebrity endorsements have included Madonna, Viktor and Rolf, Stella McCartney and Roberto Cavalli. H&M have even won the support of the pop princess herself, Kylie Mingoue, whose clothing range was launched in Shanghai in 2007. Alongside this, H&M is also a big supporter of UN programmes, the Designers Against Aids initiative and campaigns to increase the use of organic cotton, reduce the use of chemicals and the implement more energy efficient transportation. H&M is the undeniable leader in global fashion retail and looks like maintaining this position for some time to come.

SECTOR OVERVIEW

To be successful in the glossy, visceral world of high-street fashion you need to be efficient. The key for the mass-market, low-priced clothing labels is to have the season's 'must-have' in store faster and cheaper than anyone else. Fashion is still focused around three capitals: Milan, Paris and New York. Whilst London provides direction it is in New York where the power of the American buyers and of American Vogue is indisputable. High-end and high street fashion are becoming increasingly intertwined – the true fashionista mixes a thousand pound handbag with a Top Shop skirt and delights in the difference in price. To pull in the customers, the savvy high street retailer needs to get the key trends from the catwalk to the store before the next issue of Vogue has arrived on the shelves.

While the market remains relatively fragmented mergers and acquisitions as well as organic growth are starting to build global players which enable large chains with logistics strengths to compete aggressively with local brands. Fashion is no longer the province of the few, but a familiar part of the urban landscape with identikit stores across the globe. Alongside the established global brands of Gap, H&M and Zara, other companies are also growing fast. Most prominent of these have been Uniglo from Japan, Spain's Cortefiel and Hong Kong based Giordano, all of whom are expanding steadily into new markets in Russia, Europe and Asia. In addition, there has been a rise of 'own labels' within general retailers such as Wal-Mart and Tesco that have grown to gain a significant share of the markets in which they operate.

INNOVATION DRIVERS

The best high street fashion retailers have broken fashion's glass ceiling by focusing on the season's must-have items from the catwalk. These appear quickly on the rails faster and cheaper than they will be selling in the high end stores. As canny consumers mix and match to create the latest look, those retailers most adept at reading the market and their customers can keep increasing both market share and profit margins. Visual merchandising and advertising play a part in enhancing fashion credibility with the best retailers visible in the glossies, on bus stops, in catalogues and on the internet. By creating demand for clothes that look like they've come straight off the catwalk, but without the price tag, the high street is pulling in customers with significant spending power and spend per customer is growing. A number of techniques are in play in order to address this. The trend for low-cost designer lines continues with many fighting for share of voice. This has been good news for celebrity designers, or indeed celebrity fashionistas in general: on the back of H&M's lead, Kate Moss has been used at Top Shop, while fashion designer Max Azria had a new range, 'Tex by Max Azria' in French retailer Carrefour.

Several established fashion retailers are also expanding their product lines, adding items beyond clothing such as shoes, accessories and cosmetics. Franchise stores and online sales websites are also both on the increase as an effective and fast route to market. That said, the heart of high street success is now just as much efficient management as it is product design. Both Zara and H&M have streamlined the production cycle with enviably efficient production and logistics management supported by feedback and communications mechanisms within the stores that allow for the right product, in the right quantity to be sold at the right place for the right price: a killer combination.

ONES WE ARE WATCHING

Baugur

This Iceland based company continues to grow through acquisition and is now starting to make some interesting innovation-driven moves. With over 3,900 stores worldwide and a turnover approaching £10bn, Baugur's fashion brands include Nine West, Coast, Karen Millen and Oasis. With department stores also in the fold, cross-brand product synergies are high and process and business model innovation is showing promise.

Giordano

Giordano is a well extablished retailer across the Asia Pacific region and has been pushing a fast fashion mix with success for a while. With an ambition to become the best and the biggest world brand in apparel, Giordano currently operates 1700 shops in 30 territories worldwide, including over 700 stores in China alone. As it embarks on further growth, it is using new brand outlets such as Giordano Junior and Giordano Concepts to help drive expansion with gross margins of around 50%.

Changing the role of IT in the organisation by using unique technologies to innovate global business models

INFOSYS

Infosys Technologies was set up in 1981 by seven people with $250. Today it is one of India's largest IT companies with 80,500 employees covering 66 nationalities. Headquartered in state of the art facilities in Bangalore, it has offices in 23 countries and development centres in India, China, Australia, the UK, Canada and Japan. Recent times have been characterised by remarkable growth: a frequently quoted statistic, well worth mentioning, is that it took 23 years for Infosys to reach its first $1bn in revenues and only 23 months to hit $2bn. It expects sales of more than $4billion in 2008.

Infosys was the first company to successfully provide computing services from India to clients around the world, often at much lower cost. As the pioneer in IT outsourcing, it has paved the way for what has now become a $40bn industry. Today, the company handles a wide range of IT contracts for many of Europe and America's leading corporations from software development to system maintenance for multiple clients. It also provides back office support to many more, extending from order processing and customer support call centres through to high-end actuarial analysis for the insurance industry.

Infosys®

Innovation Scorecard	
Innovation Culture	9
Strategic Innovation Focus	8
New Products	7
Managed Growth	10
Product Margin	9
Investment in Innovation	8
Innovation Brand Impact	9
Innovation Peer Review	9

Infosys Key Data 2007	
Total Sales	£1.24bn
Sales Growth	35.2%
Net Income	£319m
Net Income Growth	32.5%
Net Margin	25.8%
R&D spend	£13.2m
Employees	52,700
Sales / Employee	£23.5k

Infosys employs an ever-expanding army of low-cost but highly educated Indian engineers and attracts the best of the best by offering relatively high salaries and significant perks. It is creative in the way it sources talent, drawing on the top graduates from smaller colleges. Many employees have less than 3 years experience which, given their ability and intellectual rigour, allows the company to provide its clients with a very cost-effective, yet efficient service

Some have argued that Infosys should focus on greater intellectual property creation and move beyond a conventional service model where payment is largely made for time and material inputs. They see an opportunity to move to payment for outputs, the basis of gains that their IT services can deliver. Perhaps they will: it is noticeable that Infosys's Software Engineering and Technology Labs which spearhead the company's commitment to innovation and Intellectual Property development, generated over 80 invention disclosures and filed over 20 patents in 2007.

More significantly in terms of its value chain, Infosys is moving into higher value-adding services such as establishing an R&D partnership to jointly create next-generation solutions for the power sector with ALSTOM, a global leader in power generation and rail transport infrastructure.

Its canny ability to see change and adapt to it means that Infosys is keeping ahead of the game. As it has quickly migrated from support service to value-creating software development and outsourced R&D, it has consistently been able to compete with the likes of Accenture and IBM and pre-empt new developments. For example, it has opened offices in China and Mexico to counter the wage inflation and skills shortage it is facing in India and is mitigating the possible threat of competition from other developing countries. It is also now considering acquisition and, as several other Indian firms in recent years have done, there have been rumours that it will acquire a large European firm to further underpin its future global growth ambitions.

SECTOR OVERVIEW

Globalisation has transformed the IT sector with more and more companies following India's lead and offering low-cost, high-value outsourcing services. Even the established US-based firms are now realising the opportunities from out-sourcing and off-shoring. As such, Infosys and other Indian IT service providers such as Tata Consulting Services and Wipro now face high-powered competition from the likes of IBM, EDS, CSC and Accenture as they all establish their own initiatives and add tens of thousands of new hires in India, China, Eastern Europe, Russia, Latin America and elsewhere.

However this is a challenging market where credentials and track record are key and India, with a 20 year head start on the others, is playing the pivotal role. Revenue among the top 100 outsourcing companies grew by nearly 20% in 2006 to more than $170bn. Given its size and government commitment, in the future the importance of countries such as China cannot be ignored but it will take a while for them to catch up with the lead that has been established in Bangalore.

However, as competition intensifies in the increasingly flat world that has been created from globalisation of IT services, so also does the prospect of a period of global merger and acquisition. Companies like IBM, EDS and Accenture are all wooing Indian vendors and many of these firms are more profitable than the multinational companies. So many multinational players have all invested in India to lower their costs and raise their margins. The Indian firms, meanwhile, are making selective acquisitions and are about to open large development centres in the US to compete more directly with the multinationals. To many on the outside of this increasingly complex sector, it seems the out-sourcers are now themselves doing the out-sourcing.

INNOVATION DRIVERS

The critical success factor in this sector will continue to be the ability of companies to offer a guaranteed level of service for a competitive price. A growing desire for customers to fix their budgets for up to 10 years means that IT directors worldwide are now looking to put more of their long-term contracts out to tender. Where previously these large contracts were contested only by the big western players, they are increasingly being won by the Indian firms able to back up their cost effectiveness with levels of quality, customer service and security. That said, escalating wages in India and an appreciating Rupee are putting pressure on the margins of IT companies, forcing them to move away from the traditional 'body shop' model, which was mainly dependent on cheap labour for profits, to software-assisted services which retain ownership of any software developed and charge the client on a pay-per-use basis.

As for innovation per se, few see any of the firms being particularly innovative in the traditional sense, but the way that these firms use the range of HR tools at their disposal to recruit and retain the best employees will drive innovation in working practices and HR management. However, within some of the leading firms such as Infosys and Genpact, the recent establishment of new process innovation labs indicates significant interest in modifying the business models through which such firms apply to the management of some long-term, more risky contracts.

ONES WE ARE WATCHING

IBM

IBM Global Services has long held a reputation for innovation, but has so far been unable to match the agility of Infosys in translating this into growth and profitability. With revenues of over $48bn, it dwarfs Infosys but revenue growth and net margins are lower. However, IBM is expanding its presence in India to help its delivery model to become competitive over time. If the competitive dynamics of the sector come to favour firms which create intellectual property rather than just have low cost resources, given its historically high levels of investment in R&D, IBM may gain further advantages The key challenge for IBM is therefore to leverage this investment in IP into operational performance.

Accenture

A similar story applies to Accenture. With revenues of over $21bn and revenue growth in 2007 of 18% it is larger and potentially growing as fast as Infosys but net income is only around 5 to 6%. However, with Accenture Technology Labs now up and running in the US, France and India, and one of the largest consulting arms in the sector, the company now has a significant focus on innovation and is expanding and enhancing its Global Delivery Network by, among other things, increasing activities in systems integration, application outsourcing, business process outsourcing and technology consulting areas, opening new facilities and recruiting actively in key locations.

Leader of enabling fan-based innovation to drive both incremental and significant new product development

LEGO

Why limit yourself to your own resources when you can get your enthusiastic customers to design new products for you? Over the past few years, with great success, LEGO has been challenging thinking on how to engage fans across all ages in innovation. The name comes from the Danish words meaning 'play well' which is, according the company, its ideal. Having grown from a carpenter's workshop to become the world's fifth largest toys and games manufacturer, LEGO has recently been on an interesting journey. With the deserved award of 'Toy of the 20th Century' from Fortune magazine, its core bricks, of which the company now produces 2m an hour, have become part of many of our childhoods. The company also established the LEGOLAND parks. LEGO thrived until the start of the century when it was hit hard by new technology-enabled toys and games taking significant market attention.

Innovation Scorecard

Innovation Scorecard	
Innovation Culture	8
Strategic Innovation Focus	9
New Products	9
Managed Growth	7
Product Margin	8
Investment in Innovation	7
Innovation Brand Impact	10
Innovation Peer Review	9

Lego Key Data 2007

Lego Key Data 2007	
Total Revenue	DKK8.2bn
Revenue Growth	2.9%
Operating Margin	18.1%
Gross Profit	DKK1.4bn
Net Profit	DKK1.0bn
Brick Production	19bn / yr
# Bricks / Global Pop	62
Revenue / Employee	DKK1.7m

Unable to compete for children's attention in a world of videogames and PlayStation, LEGO had to transform itself into a new streamlined global company. Alongside selling off a 70% share of the LEGOLAND parks, laying off 1200 people and outsourcing some of its manufacturing, it has used the internet to change the way it develops its products. Since 1984, LEGO had been working with MIT's MediaLab to integrate play into both the real and the emerging virtual worlds. By the end of the 1990s intelligence became an integral feature of the wider LEGO product range and simple robot technology was introduced into the portfolio.

Embracing the internet, in 2002 LEGO launched LEGO Direct and then LEGO Digital Designer which sowed the seeds for the largest children's membership community in the world. By 2004 Lego.com had over 5m visits a month as children from all over the world started to share their creations. LEGOfactory.com was launched and grew into a full online design and customisation channel that allows fans to compete against each other to design the next LEGO kit. This channel now produces many of the thousand's of customer designed LEGO sets introduced each year and the company has pushed much of its incremental innovation activity into the community.

One feature of this community is the 40,000 'Adult Fans of LEGO' who now account for about 5% of LEGO's sales. It was with this group that, in 2006, LEGO co-developed a new version of its MINDSTORMS product and, in so doing, put its' customer centric approach firmly on the global innovation map. One hundred external software engineers, teachers, architects and engineers worked virtually with LEGO to create a new product that was a step forward for the market. MINDSTORMS NXT has been a huge hit and has grabbed much media coverage, even making the front of Wired magazine.

In the last year LEGO has continued to innovate with new products and online services. Alongside the evolution of product ranges linked to franchises such as Star Wars, Harry Potter and Batman, it has introduced new versions of its own BIONICLE range of fantasy underwater inhabitants that can be controlled by the Ninetendo Wii. LEGO has transformed the way it innovates. It has become the leading example of co-development with customers and is now launching over 300,000 different LEGO sets each year.

SECTOR OVERVIEW

The toys and games market is dominated by American children who receive around $400 each of new products a year compared to an average $34 elsewhere. Split into three primary subsectors - games consoles, games software and traditional toys and games – the market is growing at around 5% a year and is expected to be worth over $100bn by the end of 2010. While children have not given up on traditional toys altogether, the introduction of videogames, computers, MP3 players and online gaming has provided fierce competition to established companies for their attention. The traditional subsector is however still over half the market and is lead by the two US based majors Mattel and Hasbro with annual sales in 2007 of $5.7bn and $3.8bn respectively. With brands including everything from Barbie, Polly Pocket, Transformers and My Little Pony through to Matchbox, Tonka, Fisher-Price and Milton Bradley, these two dominate the US market and have extensive sales across many other countries to boot.

Other key players in the sector include Japan's Bandai Corporation which has a diverse portfolio ranging from products for vending machines, cards and models through to arcade video game machines and software for home and mobile use. Number 4 and 5 in the size league are MGA Entertainment and LEGO. Alongside the manufacturers, several retailers have significant market sway. In the US, the likes of Wal-Mart, Kmart and Target control over 30% of the market while specialist chains such as FAO Schwarz and Toys R Us have had intermittent influence. However with FAO filing for bankruptcy twice in 2003 and several dot.com e-tailers failing, there has been recent realignment between the channels with, for example, Amazon.com now providing shipping and order fulfilment for toysrus.com. Although with $22.1bn sales in 2007, the US is the primary market, 80% of all toy products sold worldwide are made in China. By and large, this transfer of production has been a success but, in 2007, recalls by Mattel due to concerns about unsafe paint raised concerns for the future.

INNOVATION DRIVERS

While keeping their portfolio up to date with incremental enhancements, toy companies are constantly seeking the next big hit. As such, continual product innovation is a key to survival in this sector. So, there is huge focus on spotting the new trends and quickly launching high impact products into the market in order to capture significant share of the all-important Christmas market.

Faced with incessant competition from the likes of Sony and Microsoft, many toy companies have been making significant investment in new technologies that can be incorporated into the branded product mix. In an age when most kids in the target markets have an iPod and Facebook profile, the digital world is a primary focus. Alongside LEGO, both Mattel and Hasbro jumped on the electronics bandwagon a few years back and are now launching more and more technology enabled products.

As more and more sophisticated products emerge, an associated concern is the escalating price tags - and hence the susceptibility to declining consumer spending in the US, where retail sales of toys slipped 2% in 2007.

Licensing deals are also a major source of new products, especially in the action toys area, where some of the fees paid in recent years have been substantial. A few years back Hasbro, for example, paid $600m and a 20% royalty for rights to the second series of Star Wars movies. Now things have calmed down a bit with Mattel paying $20m and a 15% royalty for Harry Potter.

Another prominent area of attention for several companies is the emerging demographic of the 'tween' market of children between 7 and 14. As the lifespan of Barbie and similar products has become shortened, new niches have been exploited by such products as the Build-A-Bear range.

Lastly, driven both by changing socio-economic demographics and cultural references, many Chinese companies who manufacture the majority of production are being encouraged to innovate themselves. Just as Lenovo and its ilk have done in other sectors, Chinese branded toys are looking to leverage domestic market scale and global capabilities to deliver top-selling products for next year.

ONES WE ARE WATCHING

Nintendo

With new products driving renewed growth for Nintendo, it is back in the core of gaming innovation. The company was recently written off as a secondary player to Sony and Microsoft whose PlayStation and Xbox products dominated the market. Now, with its Wii and DS systems, Ninetendo is having the impact it did back in the 1980's with its Donkey Kong and Gameboy products. The Wii wireless game console has changed perceptions of what interactive gaming is all about and is now outselling the Xbox and PlayStation products combined. Associated double digit growth in revenues and margins is leading to further investment in new products to strengthen the product range.

LeapFrog

California-based LeapFrog was started back in 1995 so its founder could create products that would help his son to read. Now partly controlled by Oracle CEO Larry Ellison, the company is focused on developing a range of new products. The focus is on the learn-to-read market and expansion of the portfolio targeting the 6+ children's market. First up was the Fly, a 'pentop' computer which uses digital paper and pattern decoding to track where the user writes on the page and so helps students to write. In 2007, this was followed by the launch of ClickStart, a child's computer designed to be the first pre-school experience that looks and works like a grownup PC.

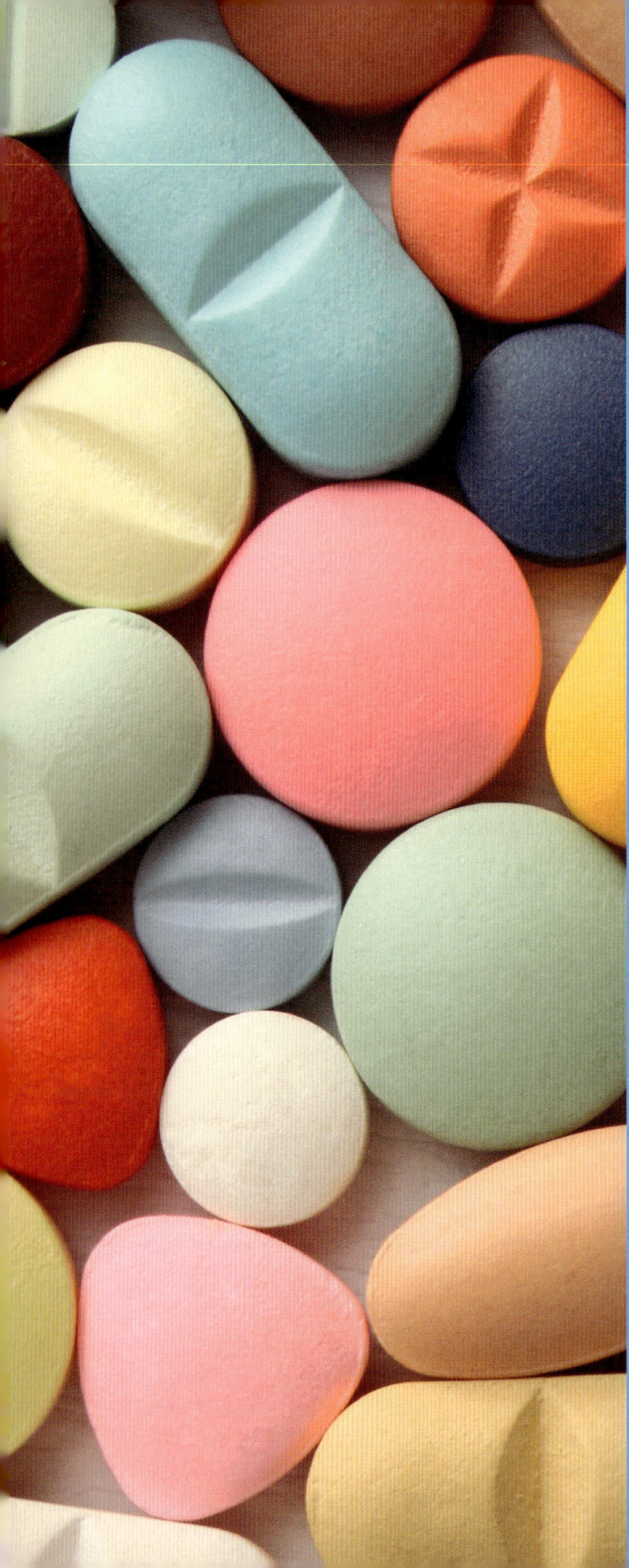

Pioneer and master of exemplary alliance management to drive sustained pipeline and portfolio growth

LILLY

From its beginning in 1876 when it was first founded in Indianapolis as a manufacturer of pharmaceuticals 'based on the best science' Lilly has been a major force in the industry. As one of the world's main pharmaceutical companies its products are now sold in over 140 countries, it has research activities around the world and Lilly is known by such brands as Prozac, Evista and Zyprexa.

With a long-term focus on the treatment of diabetes, where it pioneered injectable human insulin analogs under the Humalog brand and then moved into oral agents under the Actos name, Lilly has been widely recognised as a company that focuses its considerable \$3.5bn a year, or over \$12m a day, research effort on major challenges. Over the past few decades, it has variously successfully tackled the prevention of osteoporosis through Evista, the treatment of depression with Prozac and Cymbalta, schizophrenia with Zyprexa and pancreatic cancer with Gemzar, all of which contribute to steadily rising sales which topped \$18bn in 2007. These achievements alone do not however set Lilly apart from a number of other successful companies who have similar track records in their own areas of chosen specialisation.

Innovation Scorecard	
Innovation Culture	8
Strategic Innovation Focus	9
New Products	8
Managed Growth	9
Product Margin	8
Investment in Innovation	10
Innovation Brand Impact	8
Innovation Peer Review	9

Lilly Key Data 2007	
Net Sales	$18.7bn
Net Sales Growth	14%
Gross Margin	77.2%
Net Income	$2.95bn
Net Sales / Employee	$460k
R&D Investment	$3.5bn
R&D Intensity	18.7%
R&D Spend / Patent	$48m

What has distinguished Lilly over the past decade has been its ability to maintain successful organic growth in an age of rapid change across the life sciences sector while all around it many major competitors have chosen the route of large-scale mergers and acquisitions. While the big deals across the pharma industry have been based on increasing scale and reach, and as a result, increasing R&D effort accordingly, Lilly has led an alternative strategy. Although the company is a significant investor in its own internal research activities, it has also been at the forefront of the move to alliance-based drug development.

From its first major foray into in-licensing ten years ago, Lilly has built a significant network of relationships across the life sciences ecosystem. The company has a dedicated function, Global External R&D, focused on managing alliances with a host of universities, biotech start ups and other pharmaceutical firms. With a well-honed 'find-it /get-it / create value' process driving the securing of new technologies and molecules, Lilly has become a reference model across the industry.

In addition, Lilly was at the forefront of the Open Innovation movement, supporting the foundation, in 2001, of Innocentive, the leading marketplace for problem solving across many sectors. Within a few years of initiating its in-licensing strategy, one third of the Lilly pipeline comprised molecules that had been brought into the firm from outside and maintaining this balance across the portfolio is a key priority.

Underpinning such visible activities within Lilly there is a deep innovation culture based on building leading edge scientific capability, using venture capital to seed early stage developments and strong project teams dedicated to following up development through to post launch and line extensions. Faced with the ever-present challenges of being in the media spotlight due to the nature of its activities, the company has also been keen to be as open as possible about its research and therapeutic impact of its products. Lilly was the first pharma company to publish clinical data online – a different and unusual approach for the sector, but one that gained significant appreciation from the medical fraternity.

Going forward, Lilly is starting to undertake limited acquisitions of companies with which it has developed long terms partnerships but this complementary to its continued focus on sustained alliance-based organic growth. With 16 new molecular entities in clinical trials in 2007, and 14 new launches slated for 2008, continued sales and margin growth from innovation is expected.

SECTOR OVERVIEW

The pharmaceutical industry is one of the biggest in the world with annual sales of over $600bn. 'Big-pharma' companies such as Pfizer, GSK, Novartis, Sanofi-Aventis and Merck can and do generate multi-billion dollar profits. However, much of the industry is under increasing pressure to produce 'blockbuster' drugs that will generate annual sales of over $1bn. While sector research funding has doubled since 1991, the number of new product launches has halved over a similar period and there are understandable concerns over future profitability. Despite higher R&D investments, many of the top 10 pharma firms have been posting lower margins recently and around only 25 truly novel drugs are being launched each year. Patents are the core currency of innovation in this sector. As they expire, competition from generic drug makers is expected to threaten one fifth of the sector's annual sales. Corresponding efforts to be stronger in the use of patents to fend off generic competition leave the drug companies accused of driving up overall healthcare costs or depriving the developing world of life-saving medicines. Many in the sector embarked on large scale M&A which, in the case of GSK and Pfizer, was justified on the basis of enhanced R&D and more effective sales and marketing. Few of the mergers though, have yet to still deliver the goods and many big drug firms have started to follow the route of in-licensing more of their technology and products from outside – particularly from biotech start ups. Led by the likes of Novartis and Lilly, effectively bringing new molecules into the pipeline from outside the firm has become the current core capability in this sector. Drug discovery is increasingly undertaken outside with drug development being the primary function of the big pharma firms.

INNOVATION DRIVERS

The pharmaceutical industry estimates that out of 10,000 candidates in the lab, only 10 ever reach clinical trails and only one in five of these makes it to market. The common view is that the typical new drug development process takes between 10 and 15 years and costs $800m to $1200m for each new drug launched. With the beneficial impact of new technologies from the human genome project initially falling short of expectations, current innovation effort has been looking at how to use combinational chemistry and high-throughput screening more appropriately and find new approaches such as in-silico biology to identify the most significant targets. The key test is to unify the new technological breakthroughs with depth of prior knowledge and experience to get the right blend of skills and disciplines to deliver the much needed results.

The challenges for the sector are significant – cancer and Alzheimer's are more complex diseases than those already addressed and treatments are taking longer to develop and perfect. If potential winners can be identified earlier the ramifications will be huge. In a sector where success ratios are so low, a marginal 5 to 10% reduction in the failure rate in clinical trials would effectively double the number of new product launches. The industry wants to 'fail early and fail cheap.' US drug companies, especially, are also being squeezed by falling margins, generic competition and regulatory scrutiny and so need to improve efficiency. As well as R&D effectiveness some, such as Pfizer, have examined their sales and marketing activities. Reinventing how the products are sold is a current challenge, with reducing the size of overly large sales forces top of the list. There is also a wider argument that the sector actually needs to reinvent its business model and, like other sectors, several drug companies are looking to outsource some functions and squeeze the supply chain.

Conducting research in China and India is also an arena that has receiving much attention. With its growing pharmaceutical market, significant tax incentives, cheaper clinical trails, over 200,000 research scientists and several leading research institutions, working more closely with Chinese partners has become a priority for many major firms over the past few years. Having a research facility in China has quicklymoved from being the exception to become the norm for all the big pharma companies.

ONES WE ARE WATCHING

Genentech

Considered by some to be the founder of the biotechnology industry, Genentech has been delivering on the promise of biotechnology for more than 30 years, using human genetic information to discover, develop, manufacture and commercialise biotherapeutics. Today, Genentech is among the world's leading biotech companies, with multiple products on the market for serious or life-threatening medical conditions, and more than 100 projects in the pipeline, and is set for long term growth as it focuses first on meeting its five Horizon 2010 goals.

Merck

As it puts its recent problems behind it, Merck is again focused on using its considerable $4.9bn a year research effort to bring a range of new high-impact drugs onto the market. As well as an improved pipeline and more blockbuster products, 2007 saw growth across the board with $24.2bn of sales, an increase of 7% on the year before. Alongside this, its continued support for major initiatives such as overcoming river blindness in Africa is showing the company's ambitions for enabling life-changing impacts around the world.

Targeted acquisition of external technology to complement internal capabilities and routes to market

MEDTRONIC

In the highly fragmented world of medical devices, Minnesota based Medtronic stands out from the crowd. Founded almost 60 years ago it began as a repair company for medical equipment and quickly went on to develop the world's first 'wearable' external cardiac pacemaker. Now the world's largest medical technology company, Medtronic's operations are focused on providing therapeutic, diagnostic, and monitoring systems for cardiovascular, neurological, diabetes, spinal, and ear, nose and throat markets. It makes a wide array of electronic devices, from the implantable cardioverter-defibrillator, to devices for managing urinary incontinence and obesity. Medtronic remains focused on its original mission 'to contribute to human welfare by application of biomedical engineering in the research, design, manufacture, and sale of instruments or appliances that alleviate pain, restore health and extend life.' It does this well.

Medtronic

Alleviating Pain · Restoring Health · Extending Life

Innovation Scorecard	
Innovation Culture	8
Strategic Innovation Focus	9
New Products	9
Managed Growth	8
Product Margin	8
Investment in Innovation	9
Innovation Brand Impact	8
Innovation Peer Review	8

Medtronic Key Data 2007	
Net Sales	$12.3bn
Sales Growth	8.9%
Gross Profit	$9.1bn
Net Profit Margin	22.8%
R&D Investment	$1.24bn
US Patents	317
R&D Spend / Patent	$3.9m
Revenue / Employee	$356k

Medtronic today devotes approximately 10% of its sales to research and development efforts. To ensure that it continues to introduce innovative products and therapies, about 20% of this research budget is designated for new ventures. The company operates 26 state-of-the-art research centres around the globe and works closely with the world's leading physicians and scientists in conducting research projects and clinical evaluations of new products. These partnerships enable researchers and physicians to combine their efforts to continually evaluate and improve both new and established technologies.

Internally, a number of organisations encourage creativity and innovation by promoting the exchange of research data and technical information across the company. The Medtronic Forum is an organisation for the company's technical community; the Bakken Society, named after Medtronic's co-founder, is an honorary society that recognises employees who have significantly furthered Medtronic's technical and scientific progress. In 2007 the company published studies included reports on the removal of bacterial colonisation in chronic sinus infections and the development of minimally invasive endovascular therapies to treat abdominal aortic aneurysms.

In a bid to grow a services business, Medtronic is focusing considerable R&D effort into wireless communications. For example, in 2007 it won regulatory approval for an implantable defibrillator that links up with hospital equipment or a home monitoring device. Along with three other companies, CardioMEMS, St Jude Medical and Remon Medical Technologies, Medtronic is racing to market a device for congestive heart failure, which afflicts many millions of people worldwide. Once implanted, the device will measure pressure and fluid inside a patient's heart and wirelessly send the data to an external unit so that patients will be alerted to abnormalities at an early stage.

Medtronic aims to increase development of cell therapy and in collaboration with Genzyme, one of the worlds foremost biotechnology companies, is working to develop solutions for unmet medical needs in cardiovascular diseases through the combination of biologics and therapy delivery devices. Joint ventures such as this also work to attract leading academic, research and biotech innovators in the field to advance the treatment of heart disease. Lastly, as it builds up the scale of new acquisitions, Medtronic has bought rival spinal surgery medical devices maker Kyphon for $3.9bn in cash. Medtronic's spinal business already accounts for 21% of its revenues.

SECTOR OVERVIEW

The medical device and diagnostics industry is evolving rapidly. Demographic and societal changes have converged on the sector – chronic disease has become more prevalent, the population in the developed world is ageing, and, thanks to the internet, patients are increasingly better informed. There has been steady growth across the whole market and this is expected to continue to rise throughout the century. Accelerating globalisation and strong growth in the core emerging markets of China and India are both important factors. In the current primary markets of the US, Europe and Japan the main challenges are in the areas of healthcare economics, reimbursement strategies and new product introduction. The first two concerns are the focus of continued negotiation between the big players such as Stryker, J&J, Becton Dickenson and Zimmer, regulatory bodies and government policy makers.

The heart of these negotiations has been to agree rapid change that will help to contain expanding healthcare costs. As cost-containment spreads to the US from Europe there has also been a push to create 'world pricing' which will have a significant impact on the higher margins currently being achieved in the US, where the majority of profits in this sector are correspondingly made. Both market growth and demands for cost-control alliances and strategic relationships are playing an ever more important role here. However, it is in the last of the three identified areas, new product introduction, where companies themselves are gaining greatest differentiation and value creation. Most companies envisage future growth of around 5% annually will be delivered primarily through organic development of new products although there has also been increasing M&A activity in recent years.

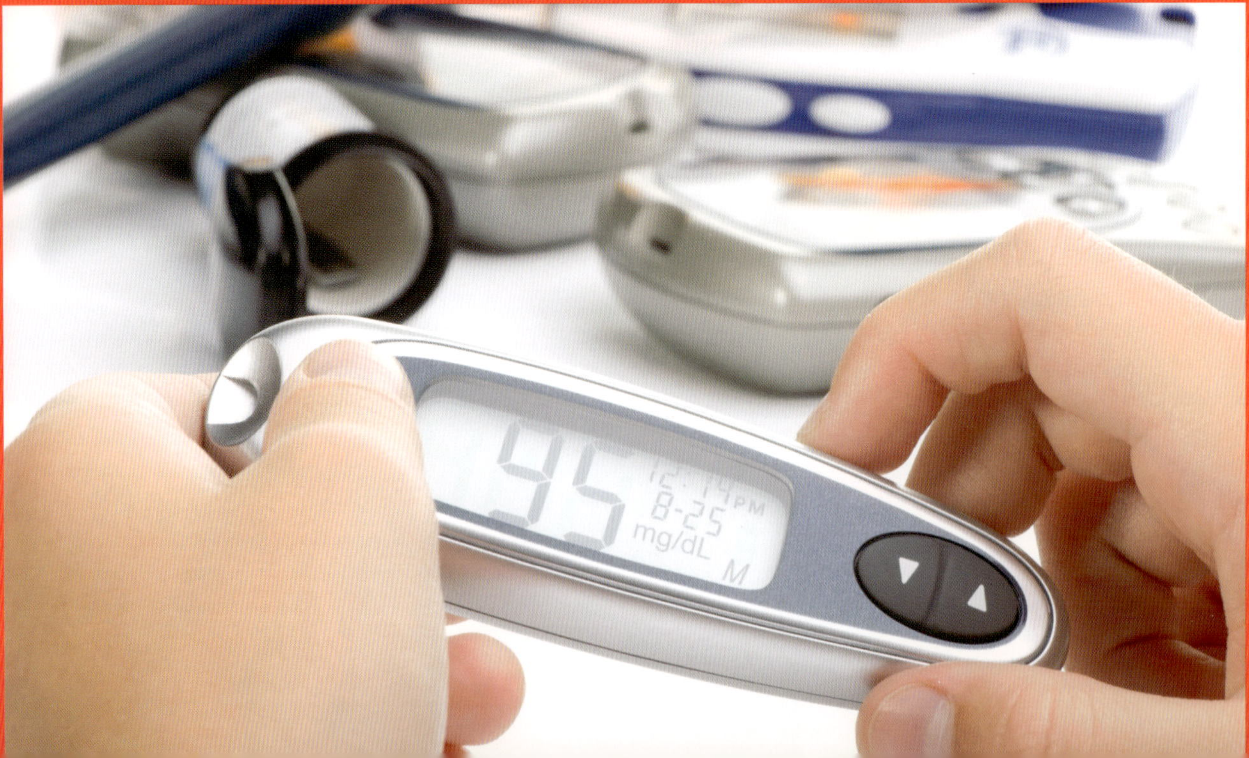

INNOVATION DRIVERS

In common with many other sectors, topics such as new product introduction, portfolio management, the voice of the customer and managing risk are all key concerns in the development and leverage of an effective innovation capability here. However, given the steadily growing demand for improved medical devices in established markets and the swift growth in emerging markets, speed-to-market is of particular priority. With the need for FDA approval, backed up by successful clinical trials, efficient decision making early on in the development process is a capability that many companies are keen to refine, particularly in the US market where increasing demand is driving prices out of the reach of many.

Innovative solutions which will reduce the burden on the overall healthcare system, while allowing growth in the sector, are seen as a matter of priority. This is now being driven as much by private healthcare insurance companies and HMOs in the US as it is by governments. All recognise the massively increasing cost of providing medical support for a growing population – some of whom are becoming, on average, healthier, but most of whom are not. As the varied impacts of smoking start to decrease, the even greater challenge of dealing with such problems as Type 2 diabetes in children as a direct consequence of rising levels of obesity in key markets is causing major concerns. Consequently innovations in products which can provide early warning of ailment, link remote monitoring to therapeutic delivery and reduce cross-contamination within the field are all gaining widespread attention. In addition, a desire for minimally invasive treatments and improved diagnostic techniques are significant areas of innovation focus.

As in other sectors, as the Indian and Chinese markets in particular expand, in-depth, accurate, predictive market knowledge to steer associated innovation priorities is also becoming critical.

ONES WE ARE WATCHING

Smith and Nephew

Smith and Nephew has been growing a niche around such areas as advanced wound management and endoscopy as well as orthopaedic reconstruction and trauma products. Having embarked on a mission to grow ahead of the market through investing in R&D and innovation, the company is now focused on building and exploiting three enabling technology platforms of novel bioresorbable polymers, tissue engineering and non-invasive stimulation across the group's business units. With sales of $2.8bn and strong poistions in all markets, the focus is now on increasing margins through new product development.

Johnson & Johnson

Johnson & Johnson is the world's most comprehensive and broadly based manufacturer of health care products, as well as a provider of related services, for the consumer, pharmaceutical, and medical devices and diagnostics markets. Always at the forefront of new developments in the sector, it has been investing heavily in creating an environment in which innovation can flourish. Recent new products include the Band-Aid liquid bandage, which creates a clear, flexible, breathable seal, and the OneTouch Ultra Blood Glucose Monitoring System, which is the first product to combine a less painful alternative to traditional fingerstick testing with the fastest test time.

Using scale and reach to simultaneously drive globally impacting innovation in the short and long term

MICROSOFT

Given its size, scale and level of success, it's unsurprising for a company like Microsoft to attract criticism. 2007 was not a particularly good year with the US Justice Department and the European Commission taking action against the company's 'monopolistic' business practices and the media sniggering about the music offering, Zune, and glitches in Vista while at the same time making ominous noises about the increase in disruptive competition from Google.

Despite this, Microsoft generated $51bn of revenue and continued to innovate, remaining a top 10 investor in R&D worldwide. It has a sector-leading track record in gaining patents, maintains sustained organic growth, employs 79,000 people in 102 countries and is now harnessing expertise and change through acquisition. If you look under the media noise, Microsoft remains one of the most significant investors in innovation in the world; one that has research teams looking out longer than many peers and one that is highly successful in monetising ideas. No surprise then that Microsoft stock is up again by 20% in 2007.

Microsoft

Innovation Scorecard	
Innovation Culture	8
Strategic Innovation Focus	9
New Products	8
Managed Growth	9
Product Margin	8
Investment in Innovation	9
Innovation Brand Impact	7
Innovation Peer Review	8

Microsoft Key Data 2007	
Total Revenue	*$51.1bn*
Revenue Growth	*21.1%*
Net Income	*$14.1bn*
Net Profit Margin	*27.5%*
R&D Investment	*$7.1bn*
US Patents	*1958*
Brand Value	*$58.7bn*
Revenue / Employee	*$647k*

There was a time when you could argue that very few innovations actually originated within Microsoft. The company's brilliance was to roll them into its operating systems and drive their popularity. The mouse, standardised tool bars and spreadsheets were all developed somewhere else first. Those 'bundling' days, however, don't stand the test of time. Changes in the way technology works mean that Microsoft is initiating a fundamental overhaul of its business practices, in order to move its existing computing platform, which currently resides on desktop PCs and servers, to the internet.

Microsoft is reacting positively to the challenge. That said, like other large companies, it has an established culture and entrenched interests which means the spirit of Open Innovation does not come naturally. There is a need to move from the era of secrecy surrounding software codes to a more open word of sociability and collaboration. The continued threat of litigation and the obvious success of Google, IBM, and Sun Microsystems in encouraging others to build software and services that fit with their own, provide strong incentives to do just this. Microsoft's new internet platform will rest on a mix of hardware, for which it doubled its data centre capacity in 2007, and software, by which it means storage and processing services delivered over the internet. In the meantime, consumers will see new services as it extends its computing platform to the internet through a 'device mesh' which will provide a way for consumers to link all of their computing devices over the internet so that personal data can be accessed from any of them.

From a strategic perspective Microsoft aims to lead and license technological solutions that become partner and sector standards. To help secure its future it leverages its corporate research labs around the world to link into emerging developments and effect fast technology transfer into the core business. Much initial focus has been on the established areas of the Microsoft empire with Vista and Office 2007. While this has met with limited success, the recently launched new X-Box game Halo 3 has, by contrast, lived up to expectations and, on its first day of sale, secured $170m in the US alone. Hits and misses are symptomatic of an innovation engine the size of Microsoft's. In other sectors the misses go unnoticed but not here. Given this, it is a notable achievement that Microsoft continues to grow in such a sustainable manner – and this is increasingly due to its hits. Microsoft's scale has not prejudiced its ability to innovate. In the arena of large company innovation, Microsoft is one of the few organisations that has proven it can deliver successfully despite its size and diversity.

SECTOR OVERVIEW

Apart from the few areas that are truly dominated by the likes of Oracle, SAP and Microsoft, the software sector is still very fragmented. As such, after years of nascent growth, it is seen by many as being ripe for consolidation and there are a number of players which are keen to assist. The intent behind these acquisition sprees is sound enough - cut central overheads and cross-sell products to a wider customer base. The reality of course is that integrating acquisitions is always a tough job: many companies underestimate just how difficult it is and find that combining several different software packages onto one platform is a time-consuming and complex operation. However as Oracle's purchase of PeopleSoft and Google's varied successes have all set the ball rolling, so others are now keen to follow.

However, for now, in this highly competitive industry, organic growth from new products is still the primary driver of performance and company valuation. Essentially separating into two distinct markets, business and consumer, there are a number of companies that currently provide software products across the spectrum. In the business sphere, the main players are firms like Oracle and SAP that focus on enterprise management, data warehousing, CRM, HR and supply-chain management. In the consumer arena, there are multiple companies ranging from the likes of EA, which is strong in gaming, through to Real Networks in media players and Dorling Kindersley in education. Amongst those working across the sector, Symantec, Adobe and Corel all have their niche and Microsoft is naturally involved in many areas.

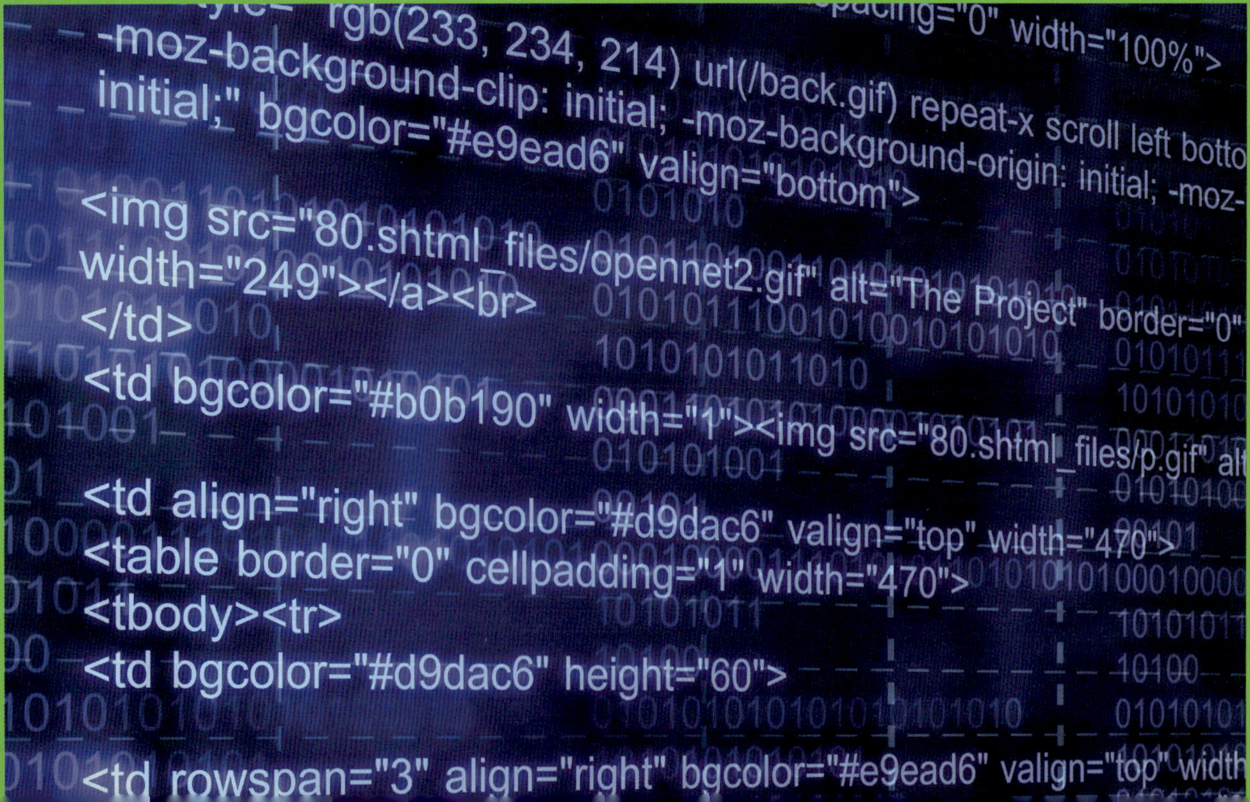

INNOVATION DRIVERS

Despite its relative immaturity, the software industry is currently reinventing itself. Wave goodbye to the days when a company was dependent on the software installed on a PC desktop and say hello to the new word processing, spreadsheets and email programs readily available to download directly from any number of suppliers including Google, Salesforce.com and Amazon Web Services. Thanks to the proliferation of broadband connections in homes and offices, customers, both consumer and business, are no longer dependent on packaged software programs which need to be installed on a computer. Instead, they can use their web browsers to tap into software supplied over the internet from central data-processing plants. Add into the mix the increasing adoption of open source and there is disruption taking place in the business applications and operating systems fields.

The next year or so will also see a scuffle amongst key players, such as Microsoft and Cisco et al as the race to provide a truly unified communications system finally kicks off for real. As the industry transforms itself from being a hardware-based proprietary technology business to a more software-based ecosystem, Cisco's advantage is that it can optimise something for the network, while whatever Microsoft develops is going to be optimised for the desktop. Regardless of which company emerges as dominant over the next 18 months, the opportunity in unified communications is large enough that both are likely to benefit. To put this in context, unified communications will account for about $4.8bn of the $40bn spent on voice, email, instant messaging and conferencing systems in 2008. This is expected to jump to $17.5bn by 2011.

ONES WE ARE WATCHING

Salesforce.com

Salesforce.com isn't a conventional software company: its model isn't about selling licenses but is based on a per user, per month charge. Focused on delivering on-demand business applications related to sales and customer relationship management, through supplying business computing as a set of simple services, the ability to disrupt the sector is clear. As companies grow by using software purchased over the grid rather than being specifically owned, they and their employees are able to control the processing of information directly, without the need for large IT support teams.

Adobe

Adobe, the pervasive digital imaging and document company, has been around for over 25 years now and has been busy building on its Photoshop / Acrobat heritage. Adobe Flash was recognised as a pivotal application in bringing animation and motion to the first decade of websites and has now developed to become the way in which video for the web has become easy to create and consume. More recently the release of Adobe AIR is spurring the next wave of innovation, bridging the dynamic capabilities of the web with the computing power of the desktop.

Leaders in high margin mass market product innovation on a journey to innovate in services to the same effect

NOKIA

As its strapline suggests, Nokia is really good at 'connecting people'. This Finnish icon is the world's largest manufacturer of mobile devices and has around 40% of the global device market. Nokia is very, very successful and, in 2006, generated revenue that for the first time was in excess of Finland's state budget. Nokia has always used innovation as a key driver for growth: first, by pioneering GSM and then by reinventing the concept of product personalisation. These days Nokia's challenge is to maintain its position in a world increasingly converged and dominated by the likes of Google and Microsoft. However, while these companies have strong brands and interesting plans for the future, they don't have control over the handset. Nokia is bundling great services with tailored, user-friendly hardware. With a billion customers and relationships with hundreds of operators around the world, Nokia may well manage to hold its place.

NOKIA
CONNECTING PEOPLE

Innovation Scorecard	
Innovation Culture	9
Strategic Innovation Focus	9
New Products	9
Managed Growth	10
Product Margin	10
Investment in Innovation	9
Innovation Brand Impact	10
Innovation Peer Review	9

Nokia Key Data 2007	
Total Revenue	€74.6bn
Revenue Growth	49.5%
Operating Profit	€11.6bn
Profit Growth	46.1%
R&D Investment	€5.6bn
US Patents	805
Brand Value	$33.7bn
Revenue / Employee	€742k

Core innovation strengths within Nokia include speed of action, interconnection between products and services and the strategic use of design. Over the past few years, as margins have been three times those of its nearest competitors, the company has re-emphasised the importance of design to place it literally at the heart of the organisation's operations. This has moved Nokia from being a feature-driven to a design-driven manufacturer where, amongst others, ethnography is now a core capability linking consumer behaviour around new uses of digital media directly into the development process. Nokia's leadership in mobile devices has given it a solid platform upon which to build a services business that expands outside the core. And here's the rub, while Nokia has consolidated its traditional strengths of R&D and product design, the big story in 2007 was its strategic shift into the multimedia services space.

Nokia created the N Series to deliver high-end multimedia phones. In addition, its near-ubiquitous camera-phones have provided revenue opportunities from user-generated content. In support of the social networking trend, Nokia has formed alliances with partners such as Yahoo to allow users to share photos using its Flickr service. 2006 saw the first tablet device with internet, not cellular connectivity, and the success of this product has led to a next-generation of internet-only devices including webcams and higher levels of VoIP. By eliminating the SIM card and breaking the connection to the networks, Nokia is using alliances with the likes of Google and Skype to offer more freedom of choice direct to consumers. At the lower end of the market, Nokia is the leading brand in China and India and is well positioned for further growth. It continues to lead on process innovation to drive down cost and is opening its tenth factory in India with the capability to turn out 20m phones a year.

Nokia's innovation roadmap weaves its software and services into a seamless package. Smart acquisitions have increasingly played a considerable role in this. Seeing that that location and content services provide major opportunities Nokia bought Navteq, the leading provider of digital map information for over $8bn in cash. Nokia expects that the truly mobile internet with multiple connectivity options that enable faster access to music, video, TV and mobile navigation and massive multiplayer gaming services will be a major factor in driving further growth. With overall industry handset volumes growing, Nokia is in the strongest position of any manufacturer. Add into this mix Nokia's acquisitions and partnerships that are building new service portfolios and it is clear that this company continues to be the leading source of innovation in the telecommunications sector.

SECTOR OVERVIEW

After a century of gradual development from telegraph to sophisticated fixed-line communication there was an initial divergence created by the newer mobile telecommunication businesses. This has been followed by increasing convergence that has brought the two core systems together, thanks to technical developments enabling cordless communication, broadband providing high-speed connection and industry consolidation from a plethora of privatisations, mergers, acquisitions and new entrants.

Nokia's leadership in the design, supply and support of a wide range of mobile equipment remains strong and, with increasing share and margins, the company has distanced itself from immediate competition. Having fought off increased challenge from Motorola which was, for a time, gaining market share, albeit at the expense of margins, Nokia is now facing more significant global growth competition from Korea's Samsung. Especially in the US where Nokia has less than 10% market share, Samsung, LG and Motorola are all trying to capitalise on their entrenched position before Nokia turns up the heat with its combined high-end product and service offers.

Despite the high penetration in most developed countries, the world market is far from mature and the next wave of handset design is currently being driven by internet access, faster downloads and a variety of connectivity options making the handset the perfect companion for communication, music, video and gaming. After success in the 'triple play' voice / data / internet arena in lead countries such as Denmark and Korea, others have followed suit, and many have jumped on this opportunity to combine mobile, landline and VoIP communication in various ways. T-mobile now offers reduced mobile rates when using your mobile phone at home, Skype has been a source of major disruption and LinkSys recently introduced a product which uses WiFi connection and VoIP independent of a PC.

Telecoms operators, handset manufacturers and software companies all feature different business models and revenue streams which make this one of the more complex industries in existence. Whether average revenue per user or margins of market share is the key metric, in the end the consumer is voting for services with price transparency, ease of use, superior design and a complete set of features matching their lifestyle requirements.

INNOVATION DRIVERS

Market share and margins matter in this industry and in recent years, from many sources these have been under pressure. As technology platforms are co-developed, shared and licensed through the likes of Symbian and Qualcomm, technical leadership no longer holds the sway it did in the past. Within the innovation portfolio, design has become the key differentiating factor and is the primary area where competing companies are continuously looking to gain an edge over their rivals. Nokia has a design process that looks for 'remarkable similarities in what global consumers want and need in their mobile devices'. While Motorola, LG and Samsung use design to focus on the medium priced segments with the RazR, the Viewty phone and the G800 respectively, Nokia has placed bets on both high-end smart-phones such as the N-95 on the one hand and low-end models to grow and maintain its market share in countries such as India and China on the other.

Alongside design and technology, collaboration with other sectors is the other major area of innovation focus right now. Because it is the networks and not the device manufacturers that own the consumer relationship, both parties are keen to innovate in new ways with new partners. Google's initial teaming up with Orange to provide a phone capable of delivering Google's services on a mobile platform was eclipsed by the November 2007 release of Android, an open source software platform for mobile devices developed in conjunction with the Open Handset Alliance. At the same time, Google is also busy cooperating with the likes of Nokia in delivering location-based advertising, and Nokia is itself leading a change into services through the 2007 launch of Nokia Music Store and a host of new developments.

Having experienced massive technology convergence within the device as handsets simultaneously act as phone, camera, computer and internet access, this sector is moving fast into a second generation of product and service convergence. For many consumers, their phone is the key item they carry around with them all the time. It is the first thing they use when they wake up and the last thing they touch before they go to sleep. With such global consumer acceptance, more and more companies are seeking to leverage this ubiquitous device to deliver an increasingly wide range of innovations from healthcare through to entertainment.

ONES WE ARE WATCHING

Sony Ericsson
Seven years after its establishment as a joint venture, a resurgent UK based Sony Ericsson is using design-led innovation to drive growth. With smart-phones such as the P990i and Walkman branded products such as W580 helping the company to achieve over 100m unit sales in 2007, margins and market share are both rising steadily. As it continues to integrate in high-end camera performance, future growth to over 10% of global market share is expected.

Qualcomm
A leader in developing and delivering innovative digital wireless communications products and services based on CDMA, Qualcomm has been using its strong intellectual property positions to capitalise on opportunities in Asia and the US. As a major rival to Nokia's alternative technologies, it is a strong partner for many of the new entrants seeking to gain stronger footholds in the global device marketplace and so is very much an enabler of future growth.

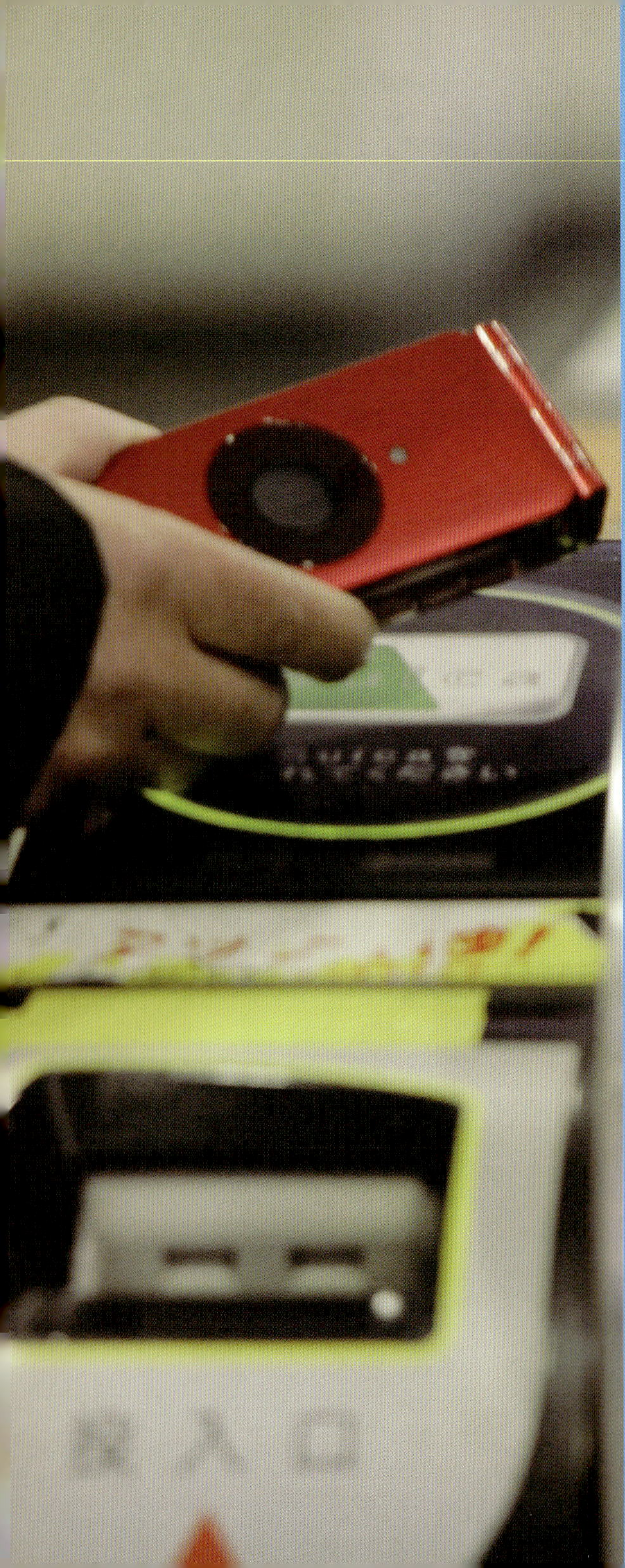

Getting on with launching new innovations into the mobile marketplace while most competitors dither

NTT DOCOMO

As the predominant mobile phone operator in Japan NTT DoCoMo is aptly named, reflecting both the Japanese 'dokomo', meaning 'everywhere' and the English phrase, 'Do Communications Over the Mobile Network.' Given that it has 54% of the Japanese market and its brand and service brands such as i-mode are synonymous with the phenomenal growth of Japan's mobile phone industry, it certainly lives up to its name. Originating as a spin-off from the incumbent national telephone company NTT in 1992 the DoCoMo business provided its first cellular phone service in the same year and has been growing ever since. It is now one of the world's largest wireless operators and has pioneered a number of major innovations. In 1999 NTT DoCoMo launched the i-mode platform to stimulate the use of mobile internet services in a flattening market for voice communication and two years later became the first global telecom operator to offer third generation (3G) mobile phone services. While over 60% of the company's customers have now switched to 3G, the i-mode platform turned out to be a greater initial success, providing a good level of functionality to a wide range of customers. As it faces an increasingly mature market, DoCoMo sees innovation as the means of maintaining its market leadership and profitability.

Innovation Scorecard

Innovation Scorecard	
Innovation Culture	7
Strategic Innovation Focus	9
New Products	8
Managed Growth	7
Product Margin	8
Investment in Innovation	7
Innovation Brand Impact	8
Innovation Peer Review	8

NTT DoCoMo Key Data 2007

NTT DoCoMo Key Data 2007	
Total Revenue	4.8tr Yen
Revenue Growth	0.5%
Gross Profit	2.8tr Yen
Net Profit Margin	9.6%
Net Income	476bn Yen
US Patents	148
Employees	22006
Revenue / Employee	2.2m Yen

In a market where talk of innovation dominates, NTT DoCoMo has been getting on with the job. Key to its success is the provision of easy-to-use advanced mobile network services such as the development of e-commerce offerings such as the 'Osaifu-Keitai' or digital wallet. This allows a mobile phone to function as a cash card, credit card, train or plain ticket. By the end of March 2008, the company had 4 million subscribers for its new credit card business and 250,000 payment terminals in convenience stores, fast food outlets and other shops. Not bad for a mobile network operator. Access to DoCoMo's customers, about 40% of Japan's population and more than half of its mobile phone users, has even convinced the traditionally conservative McDonald's to use the wallet functionality as the basis for a loyalty-based membership club. DoCoMo already counts multiple urban transit systems, convenience stores, Coca-Cola vending machines and some online shopping sites among those that offer its payment network, but McDonald's is a very high-profile addition for the company.

DoCoMo owns a stake in the network that processes and clears the transactions, so as this service gains size and scale, it offers another potential revenue generator for the company. DoCoMo also provides innovative e-commerce services for business users and has teamed up with Casio Computer Co to provide digital payment systems, mainly for its 'iD' mobile-based e-wallet services. This provides digital payment and customer relationship management services, enabling owners of shops and restaurants to manage and analyse their sales information.

In parallel with these developments NTT DoCoMo is also changing from being a technology-driven provider to an organisation that responds to consumer needs. This, despite the fact that it already boasts one of the highest customer satisfaction scores of any operator and has the lowest churn rates in the industry today. A good example of this is the i-Channel news feed which offers a keyword search service to makes it easier to find and use information. This service is free and boasts over 5m users. The development of a 4G phone, mobile advertising, location services, push-to-talk walkie talkies, child-safety phones, the provisions of games that rely more on customers moving the handset as per Nintendo's Wii, not to mention the Wellness Phone which gives health checks, measures calories, and even offers a breathalyzer are all on DoCoMo's agenda. Looking forward with a strong domestic position and eager customer base NTT DoCoMo is set to continue its role as a pioneering operator, one able to work with equipment suppliers to enable the next generation of services ahead of the pack and so set the standards for future growth.

SECTOR OVERVIEW

The telecommunications industry has had to come to terms with a change in the weather, particularly in Europe where the glory days of steadily rising mobile penetration and secondary growth through prepaid tariffs and SMS are now, if not a mere memory, certainly in steep decline. After the frantic investments in the late 1990s and the failure of 3G services to attract a mass customer base most operators are casting around for greater returns. Traditional players are further challenged as they face unforeseen competition from flat-rate wireless VoIP, WiFi and the threat of mobile broadband based on WiMaz. Some telecom and cable operators have responded by offering triple or even quadruple play, which consumers largely see as an opportunity to lower their overall communication costs.

As new entrants challenge their core operations, several telecom operators are also venturing into new businesses. Long-awaited mobile banking services are now becoming a reality – particularly in developing markets where mobile technology has not only leap-frogged existing landlines but also looks set to give the bankers a run for their money. Not without a fight however – take for example the Dutch bank Rabobank which has now become MVNO to increase the overall use of mobile banking services. In addition, with companies such as Virgin successfully linking up a compelling product mix for consumers through providing them with a full suite of mobile, broadband media and banking services, further opportunities for scale driven mergers and more interesting business model innovation are plentiful.

INNOVATION DRIVERS

The mobile industry needs to standardise technology platforms so that it can stimulate greater innovation and bring increased value to its consumers. With the rollout of new networking technologies such as 3G, 4G and interoperability with WiFi and Future WiMax, higher bandwidths will accelerate development of new services. The industry's track record in service offerings to date has been poor and, with young pretenders such as Google and Apple snapping at its heels, it can't afford another failure such as WAP or UMTS.

Traditional operators are growing through acquisition in developing markets – take for example Vodafone's recent purchase of Hutchison Essar for $11.1 billion. In the short term this enables them to return to what they are good at, the acquisition of new customers, but it has also made them focus on new products and services which are relevant to the markets in which they are operating. Mobile banking services, for example, have as yet failed to get much traction in the West, but they may well have a significant reduction on churn in markets such as Kenya where there is little or no banking alternative. Expect also the delivery of more new low-cost, low-functionality handsets as operators go back to the basics.

Concurrent to establishing new markets operators must ride the Web 2.0 wave and become internet-based businesses. Despite the attendant risks to privacy, social networking sites, specially adapted to mobile, will soon be the norm; charging models will adapt to mobile advertising and reduced rate or even free calls will be available. Gaming, music downloads and Mobile TV are all set to become more sophisticated, localised and interactive. The common innovation driver behind all this? Google.

ONES WE ARE WATCHING

Vodafone

The world's largest mobile phone company has for many years been focused on growth through acquisition of customers. However as markets near saturation and its India investments gives it a strong emergent market, the company is increasingly looking for new platforms to drive growth in both existing and new areas. Most notable in recent years has been M-Pesa, the disruptive, low-cost, mobile payment system, that is having a major impact in Africa and beyond. With this success under its belt, the company is becoming increasingly proactive in searching out new major opportunities.

Korea Telecom

Like NTT DoCoMo, Korea Telecom has benefited from close domestic partnerships which have enabled it to be a pioneer of advances such as Ultrabroadband and Wi-Max in major urban areas. With a strong relationship with fellow Korean firms Samsung and LG, Korea Telecom has been able to be the platform for much of their new technology developments from mobile internet and TV through to mass gaming. Now expanding strongly outside its home market, including some interesting moves into Russia, Korean Telecom is an emerging force in global communications.

Proactive refocusing of product portfolio through innovating around defined strategic priorities for the future

PEPSICO

Many of us know what is 'good for us' but we never quite manage to make the necessary lifestyle changes. Multinational companies are broadly similar. The exception is PepsiCo. With a growing consumer demand for healthy products, PepsiCo has cleverly managed to bridge the potentially conflicting trends of health and convenience. The company still offers 'fun for you' foods but early on it recognised the demand for healthier products and took action. So much so that it now commands half the US market in 'good for you' food and drinks; whatever else, this has certainly been good for PepsiCo. Formed in 1965 by a merger between Pepsi-Cola and Frito-Lay, PepsiCo has become the sixth ranked food and drink company in the world. The 'fun for you' snacks are no longer the main focus for the company. The interest is now clearly in 'better for you' and 'good for you' products. As change goes this has certainly had impact.

PEPSICO INTERNATIONAL

Innovation Scorecard	
Innovation Culture	8
Strategic Innovation Focus	9
New Products	8
Managed Growth	9
Product Margin	8
Investment in Innovation	7
Innovation Brand Impact	9
Innovation Peer Review	8

Pepsico Key Data 2007	
Total Revenues	$39.5bn
Revenue Growth	12.0%
Gross Profit	$21.4bn
Profit Increase	10.0%
Net Profit Margin	14.3%
# $1bn Brands	17
Employees	185000
Revenue / Employee	$211k

Back in the mid 1990s and well ahead of its rivals, PepsiCo saw the future threat that increasing levels of obesity would play and decided to change its product portfolio. In 1997 with saturation in the fast-food marketplace the company sold its Pizza Hut business and made its first major acquisition – that of the fresh fruit juice company Tropicana which was swiftly followed by Quaker Oats for $14bn. The last five years have seen a major drive to remove hydrogenated fats from many existing snack products, coupled with a pledge that 50% of new products will use, where possible, 'healthy' ingredients or offer health benefits. This policy has even led to the production of fat friendly crisps; for example, the Walkers Potato Head crisps have 70% less saturated fat than traditional products.

Continuous innovation in snacks foods is not the only focus. In the beverage market the decline in carbonated drinks sales has been more than offset by other products. PepsiCo's portfolio no longer depends on the Pepsi, 7Up and Mountain Dew lines with which it made its name. The company's Aquafina bottled water is the number 1 brand in the US and its Gatorade sports drink now has 80% of the market. In addition, through its joint venture with Starbucks, PepsiCo now dominates the bottled coffee market with the Frappuccino range of products.

Across the product range there has also been increased recognition that 'international', or perhaps more accurately non North American, businesses are key to the future. After success in emerging markets such as Brazil, China, Russia and India, PepsiCo expects its international businesses to grow at twice the rate of that of North America. The company recognises and acts on the need to adapt to cultural differences in taste and style and consequently in the future the core portfolio may not be the main driver. Take Russia, for example, where the fizzy drink craze of the west has not really caught on: fully half of PepsiCo's Russian beverage business is noncarbonated drinks – juice, water tea and energy drinks.

From being a company always under the shadow of Coca-Cola, PepsiCo has moved to the front and is intent on staying there. Since 2003 the company's stock has more than doubled as it has out manoeuvred Coca-Cola, the fall of which over the past five years has mirrored PepsiCo's rise. With over half its revenues coming from increasingly low fat foods and an ongoing change towards healthier drinks products, the company is the proactive food and drinks company, willing to cannibalise its own portfolio and best position itself for long-term growth.

SECTOR OVERVIEW

In many countries, the food and drink market is mature and competitive. The structure of the sector has been consolidating, as major companies use economy of scale to gain market share, and there has also been a strong trend towards internationalisation with acquisitions playing a major role here. Whilst global brands and global foods still largely dominate the market, these are being blended with local preferences to meet the growing demand for local products. Consumer purchases in developed countries are driven by convenience, health and pleasure. Desires for low carbohydrate and organic foods have hit the industry over recent years with corresponding changes in many brands' portfolios. While the shift towards organic has been important in some markets, globally there is increased mainstreaming of natural foods.

There is additional pressure due to increased regulation in food safety, health and traceability and the inherent cost of innovation as well as price erosion of the final product, coupled with the availability and increasing price of some key raw materials. Of great concern in 2008 is the escalating cost of grain which is being driven by the thirst for biofuels in the US. Not only is this impacting the dynamics of much of the food sector from tortillas in Mexico to food ingredients in Europe but it is also increasing the pressure of key areas of the drinks markets – especially beer production.

At the same time the rise of obesity from over-consumption is showing little signs of slowing. However, with nearly two billion people now significantly overweight, 800 million still go hungry every day. The great challenge here is that there is no overall problem of world food supply, more one of food distribution. Food is being produced in growth markets to feed increasingly fat people and economic and regulatory barrriers are still a major concern in many countries. Added to this continued global population rise and migration are adding to the challenge. As the huge corporations such as PepsiCo, Nestlé and Unilever that dominate the sector all seek to change their developed world portfolios, there are significant unmet needs elsewhere.

INNOVATION DRIVERS

Innovation in the food and drink sectors is generally driven by trends in diet and health, technology, farming and the environment, plus demographic and social change. However, far and away the sector's biggest driver for change over the past few years has been globalisation: the international integration of markets means that cross-border trade is increasing and retail has become a global business. Global, multi-market foods and drinks now account for the majority of purchases and so have become a major area of focus as companies seek to accommodate such mega trends as Asianisation of the product mix.

In terms of food consumption, thanks to the move from traditional meal times to snacking and dashboard dining, food is increasingly available through non-traditional convenience outlets such as petrol stations, kiosks and vending machines. The net result is a growing range of foods that can be prepared in less than 15 minutes and 'snacked'.

At the same time, the widely evident sustainability agenda is now driving concern across the board from reducing packaging and lowering distribution mileage through to limiting water consumption and reducing carbon footprints. As 'doing the right thing' has become a mainstream issue for middle class consumers around the world, food and drink companies have had to respond quickly to this major shift. Sustainability now ranks alongside lifestyle and health as a significant driver of consumer interest.

Although taste is still the primary factor in most food and drink choice, nutrition is rising fast. As health concerns increase, companies have been quick to seize on the demand for functional food and drink with products, such as probiotics and prebiotics, that either have some specific nutritional or dietary benefit or can be targeted as such, being king. The buzzword here is neutraceuticals which can be found in most major companies' innovation strategies.

Lastly, there is also strong focus on improving the core production of foods and drinks by including new biotechnology and genetic modification to develop better texture and flavour, longer shelf life and easier shipment, better yield and higher nutrient value.

ONES WE ARE WATCHING

Nestlé

The Swiss based food giant Nestlé used to struggle to deliver sustained growth but over the past few years this has ceased to be a problem. Fuelled by an increasing innovation focus on health and nutrition as well as major investments on luxury and premium products, Nestlé has achieved significant benefit. In 2007, its twelfth consecutive year of growth, profits were up 15.8% with overall sales up 9.2%.

Cargill

Cargill is an unknown giant in the food sector. Privately owned and hence off many commentators' radars, this vertically-integrated US-based firm is a major player across the sector. With significant growth now being delivered from its biofuels interests, its core strengths in science-led innovation around such areas as grain development are enabling it to help address the key global challenges of producing high quality food for all.

HOUSEHOLD GOODS

Obsessive delivery of consumer-driven, margin-increasing and fast incremental innovation across all markets

RECKITT BENCKISER

Reckitt Benckiser, the world's #1 household cleaning products firm, came into being as the result of a 1999 merger between the UK's Reckitt & Colman and the Dutch group Benckiser. The company leads the sector in the manufacture and sales of household cleaning products with sales increasing on average by 7% a year since its formation and a share price rise of 356% - compared with a 13% decline in the FTSE 100 index. From its European origins it continues to expand globally with its brands now available in more than 180 countries. Nearly half of its revenues now come from outside its home markets, with the US and Australia accounting for 28% of net revenues and developing markets 18%.

Reckitt Benckiser is passionate about delivering innovation. In fact, 40% of net revenues are from products which are less than 3 years old. Over the last year new products were introduced across major brands with numerous line-extensions and regional roll-outs maintaining momentum. The associated growth across the business is outpacing much of the industry and, with the establishment of what it calls its Power Brands, the company has achieved number one or two positions in 75% of its markets.

Innovation Scorecard

Innovation Scorecard	
Innovation Culture	9
Strategic Innovation Focus	9
New Products	8
Managed Growth	10
Product Margin	8
Investment in Innovation	9
Innovation Brand Impact	8
Innovation Peer Review	8

Reckitt Benckiser Key Data 2007

Reckitt Benckiser Key Data 2007	
Total Revenues	£5.3bn
Revenue Growth	7.0%
Operating Profit	£1.2bn
Profit Increase	35%
Operating Margin	22.6%
Gross Margin	58.3%
Media Investment	12.4%
Revenue / Employee	£239k

Alongside the incessant organic growth, 2006/2007 saw the first major acquisition for many years with the £1.9bn investment in Boots Healthcare International (BHI). The logic behind this acquisition was the assumption that an innovation focused consumer goods firm is more able to successfully support the retail of health care products than pharmaceutical companies. Healthcare is acknowledged to have a steadier, longer-term innovation cycle driven by medical claims and regulatory approvals rather than rapid new product launches. The acquisition has provided Reckitt Benckiser a new platform for growth and margin improvement through products like Stepsils, Nurofen, and Clearasil which are all now performing above expectations. Following the success of the BHI acquisition, at the end of 2007 Adams Respiratory Therapeutics, which owns cough and congestion drugs Mucinex and Delsym, was also brought into the fold to further add to the OTC portfolio, giving Reckitt a stronger foothold in the US.

Driving all this growth is a company that has innovation as its #1 objective. In a sector where it really is a case of innovate or die, from the CEO down, everyone is motivated and rewarded by contribution to innovation driven growth. Reckitt Benckiser's culture is one which encourages innovation through a combination of limited bureaucracy, ambitious product targets and performance based reward structures. This system clearly works as multiple products are launched in different markets and the winners are spotted quickly. Successful new launches such as that of Cillit Bang in Hungary are spotted within weeks and then scaled up for global launch in a couple of months. While competitors struggle to find the next big thing to drive growth, Reckitt Benckiser gets on with launching numerous incremental innovations, from which it selects the high performers and then rapidly builds and extracts value.

Alongside its culture, another major differentiator of Reckitt Benckiser against many of its peers is the way that consumer insight, rather than high R&D investment drives product development. The company works hard to identify new cross-sector trends and highlight new opportunities for improved supplier-led innovation and employees across the business are encouraged to challenge traditional assumptions and approaches. Having raised operating margin from 14.4% to 22.6% over the past six years and achieved double the like-for-like sales growth of the sector's two largest companies in 2007, Reckitt Benckiser has now set a target of 7% sales growth and 10% profit growth going forward. At this rate, there are few in the industry who would doubt that this innovation powerhouse will make it.

SECTOR OVERVIEW

The household goods sector is an amalgam of several product groups including personal care, oral care, laundry, paper products and household cleaning. It is dominated by a number of multinationals such as Procter and Gamble, Unilever, L'Oreal, Kimberley Clark, Henkel, Colgate-Palmolive, Reckitt Benckiser and Beiersdorf. Some pharmaceutical corporations also have an involvement in personal and oral care and there are specialist players including SC Johnson. In addition, retailers, and particularly Wal-Mart, are beginning to penetrate this sizeable market space with their own label value offerings: unsurprising when you consider that the US oral care market alone is worth c. $7.4bn and the household cleaning market is now touching $4.5bn. With its 8% share of all US retail shopping, Wal-Mart's buying power is strong.

Within highly competitive markets, brand association combined with innovative products are the key to commanding the added value sales. For example the global brands of Colgate and P&G's Crest lead the oral care market; Beiersdorf's Nivea, Unilever's Dove and P&G's Olay are all strong in personal care, while Reckitt Benckiser's Dettol, Vanish and Lysol make it the leader in household cleaning. Across the sector, the core trends influencing consumers' purchase decisions include convenience, fragrance, confidence and product effectiveness. It will also be interesting to see the impact that concerns around the increase in viruses, particularly avian flu, will have on the growth of antibacterial products. In addition, as with many sectors increased consumer environmental awareness has stimulated a demand for less water use, less packaging and more eco or natural product variants.

Leading players are also keeping their eye on the growing market for low-cost detergents, toilet tissues and soaps available through discount stores. So far the impact on market shares has been marginal but increasing declines in consumer confidence and fears of a recession might change this. Finally, with the overall household products market mature in many regions and much future growth now being targeted at China, Eastern Europe and SE Asia, delivering successful differentiating innovation is seen by all key players as a primary concern.

INNOVATION DRIVERS

Geography, demographics and consumer economics all influence growth in this broad sector. In personal care there is a strong trading-up focus for incremental innovation as increased consumer demand for convenience in emerging markets has, for instance, grown the shower gel market at the expense of the bar of soap. At the other extreme, areas like oral care have had a period of more radical innovation. Strip breath fresheners, electric toothbrushes, whitening toothpaste and tooth gels have transformed this previously slow-moving market into a leading innovation area. Across the whole sector, there are several key drivers having a tangible impact on how and where companies focus their innovation activities. Customer loyalty to some products is relatively low, so brand-switching is a significant issue. To combat this, companies are all increasing their focus on successfully engaging consumers and gaining a better understanding of their attitudes. Fragrance has emerged as a key issue, propelling growth in the air care market, but also in areas such as surface cleaners, detergents and laundry products. Several organisations have increased the depth of their partnerships with fragrance experts to put them ahead of the pack in this area. That said, convenience is still the largest innovation driver here - especially given continued consumer willingness to pay premium prices for products that promise quicker and easier use. This is impacting new product development, spurring the creation of new 'system' products, such as wipes, and leading to the desire for improved integration of applicator and packaging design into the innovation process. Most recently the growth of task specific solutions has been superseded by products offering all-in-one benefits that help to simplify cleaning decisions. Looking forward, "conscientious consumerism" is on the rise. European consumers in particular are increasingly shopping with their conscience, with rapid growth in sales of organic, local, humane, Fair-Trade, and eco-friendly goods. Following the food and drink sector, personal care and household items are now responding to the commercial potential. In addition, nano-technology is influencing significant changes. Products in development cover a wide area including nano-particulate controlled release systems, porous nanoparticles for encapsulation applications and nanofibres in cosmetics. Developments are expected in hair care, body odour elimination and packaging as well as new variations of household cleaners, antibacterial agents and cleaning materials.

ONES WE ARE WATCHING

P&G

P&G, now including Gillette, continues to lead the personal and household sector in terms of scale. It has set the standard for improving internal consumer understanding, time-to-market and strategic use of design, and is also using its well-established Connect-and-Develop Open Innovation programme to increase revenues from externally sourced ideas. It has also embarked on a regional growth plan with, for example, Autodish entering the European stronghold of Reckitt Benckiser.

Kao

Kao, the 8th largest company in the personal and household category, is a rising star from Japan and Asia. With consistent growth over many years the company has recently re-organised with a greater consumer focus and a drive for new high-value-added products. Despite rising raw material prices and strong competition, sales increased by over 26% in the last fiscal year to March 2007 due largely to the success of its premium brands Jergens, Biore, Kanebo and Molton Brown.

Using design-led product innovation and excellence in technology to exploit the global value opportunity

SAMSUNG

Samsung Electronics is part of one of the largest multi-billion dollar corporations in the world. In 2007 it exceeded the $100bn mark in annual sales for the first time in its history. This makes it one of the world's top three companies in the electronics industry where only two other companies, Siemens and Hewlett-Packard, have posted larger revenues. The name Samsung literally means 'three stars' or 'tristar' in Korean, reflecting the Samsung Group's dominance in two further sectors: Samsung Heavy Industries and Samsung Engineering and Construction. If you are talking innovation in Samsung walks the walk and is now the established leader in consumer electronics, providing a range of leading-edge premium products and, in their own words, 'leading the digital convergence revolution'. In so doing Samsung has made a remarkable transformation from copy-cat manufacturer to become Asia's most valuable technology company.

Innovation Scorecard	
Innovation Culture	8
Strategic Innovation Focus	9
New Products	9
Managed Growth	8
Product Margin	7
Investment in Innovation	10
Innovation Brand Impact	10
Innovation Peer Review	9

Samsung Key Data 2007	
Net Revenues	KRW63.2tr
Net Income	KRW5.9tr
Operating Profit Margin	9.4%
Brand Value	$16.9bn
R&D Investment	KRW4.0tr
R&D Intensity	15.8%
# US Patents	3362
Revenue / Employee	KRW456m

Samsung today owes much of its success to its Value Innovation Programme. With 6 design labs staffed by 450 people it is serious about understanding what it is consumers need long before considering the technologies required to deliver them. It believes (and, to be fair, evidence suggests) that success in consumer electronics can only ever be short term and there is therefore a pressing need for continuous innovation in order to develop new technology platforms and create products that are first of its kind in the marketplace.

Samsung spends more than $6bn on research annually. It recognises that many of its products, such as semiconductors and flat-screens, are now basic commodities, and its focus is on producing iconic devices for the next generation as Sony's Walkman was in the '80s and the iPod is today. Samsung's innovation focus is therefore set firmly on design and, most specifically, on the design of digital TVs. Samsung launched the R7 LCD TV in 2005 which paved the way for the 2007 "Bordeaux", a flat screen television with contours reminiscent of a wine glass. This is Samsung's first LCD television to sell more than 1m units.

Samsung has also turned its hand to designing slick mobile phones, teaming up with Bang & Olufsen to produce the Serene, and most recently the Serenata handset. Described by FHM as "cooler than an Eskimo in an Armani anorak" it is certainly giving its peers a run for their money. In the third quarter of 2007 Samsung's mobile phone division overtook Motorola to gain second place in the market and has again seen growth in handset sales reaching 14% share. Although still far behind Nokia, with the increased demand for 3G handsets and its strong position in emerging markets, Samsung expects a further growth.

However, it is not all flash and glamour: Up until recently Samsung's memory division was responsible for 70% of the profit but overcapacity has led to an industry wide decline. Undaunted however, Samsung, which often makes big investments during downturns so that it can increase its market share and make bigger profits when the industry cycle picks up, is raising its capital investment in this division. As peers cut back this is a bold move that signals a strong confidence in the eventual upturn of the memory market. With a brand value now greater than Sony, it is clear that Samsung's strategy of delivering high quality products with an emphasis on design is paying off and the company is confident of its future performance. Time and again Samsung has proved it is able to look into the future and create what's just around the corner for the rest of us.

SECTOR OVERVIEW

In the past decade or so, the number of players in consumer electronics has grown exponentially. New markets have created opportunities for new brands and low-cost manufacturing has broken free from the low quality stigma. As such, traditional leaders such as Philips, Sony, and Matsushita have been joined by a wide range of newcomers such as LG and Samsung. A period of consolidation in the 1980s was driven by an economy-of-scale rationale. This no longer applies. There are numerous high quality component suppliers for DVD players, TVs, radios and digital cameras which now feed a global market; the barriers to entry as a branded source of products have never been lower.

Alongside the major international brands such as Samsung and Sony, hundreds of other makes are found in every high street, supermarket or on-line retailer. Throw ever faster technology development into the mix, product introduction and standardisation have both accelerated. Many of the major players in the sector now recognise design as a major differentiator alongside superior technology development and have been increasingly embracing leading-edge design into their core. Sony and Philips have always had strong design teams but now they are equally matched by LG and Samsung internal resources as well as by the multiple design agencies used in this arena.

Despite being increasingly competitive and price conscious, with consumers eager to get their hands on the latest iconic gizmo, this is a market in which many firms are keen to participate and so is one in which more and more Asian firms will enter. This is particularly true for those in China which have been long-term production centres and, like their Taiwanese counterparts, are now moving up the value chain.

INNOVATION DRIVERS

The consumer electronics sector is one which clearly displays the volume versus margin strategic option. Many brands target the highest possible volumes for their low-cost, low-margin products, while others, Bang & Olufsen and Bose being prime examples, focus on achieving the highest possible margins on their relatively low volume sales. Common to both is a need to access the latest innovative technologies – be they automated production techniques or leading edge materials.

It's true that the industry faces a range of new challenges such as network convergence, broadcasting telecoms integration and robotics but strangely the war for consumer electronics domination is not about technology. Televisions, mobile phone, and microwaves are all increasingly considered to be household items rather than "gadgets". With this in mind it is design that is the differentiator – which is why places like Taiwan and China, in the past the home of product assembly, are now developing and designing their own products thus adding to the competitive pressure.

As a result brand positioning and recognition is becoming increasingly important, particularly for those companies whose product ranges span the entire consumer electronic spectrum. Samsung, Casio and Nokia have concentrated on building their own brands while others have an alternative strategy and partner with other leading brands to provide innovative joint venture products that connect better with the target consumers. Philips is a good example of this and has teamed up with Alessi for kitchen appliances, InBev for PerfectDraft and, most successfully, Douwe Egberts for the Senseo coffee machine.

The final major area of innovation focus is the convergence between multiple devices and content delivery. Here the push for standardisation and partnership across many different sectors ranging from telecom, banking, retail, consumer products, IT and many others is becoming vital for the success of this industry and its strive to further advance in the home and office environments.

ONES WE ARE WATCHING

LG

LG is on a similar path as Samsung going for creating a stronger brand and delivering top quality products in the home products arena. Its joint venture with Philips has helped establish the LCD TV portfolio and, as demonstrated by recent launches such as the Prada and chocolate phones, this traditional me-too innovator is moving fast up the innovation value chain and seeking to exploit its broad technology and product portfolios.

Sony

Although making significant losses in recent years, Sony's consumer electronics group is fighting back and with the recent win in the battle of the high definition DVD formats going its way is set to capitalise on broader opportunities to use this as a platform for broader innovation. In addition as its Vaio range of computers and the ubiquitous Walkman gain strength, some would see a positive future for this previous star of the sector.

Identifying, understanding and seeding the development of future growth opportunities ahead of its peers

SHELL

Every day more than 20 million customers in 130 different countries visit Shell service stations for fuels, motor oils, car care products and the odd sandwich or two. In the face of fast rising global demand for energy, like many of its peers, the company has been using the latest technological solutions to both discover and produce oil and gas, process and refine products, and also develop energy from new sources. However, unlike many of its competitors, Shell has also demonstrated the benefits of its focus on leading the development of future opportunities beyond the traditional energy space. Shell's 2007 results recorded the largest profit ever reported by a listed European company, and, despite some media accusations of obscene profits, the reality is that as fast as Shell pulls in profit, it is reinvesting it to prepare for the future. The group's capital expenditure for 2007, excluding acquisitions and disposals, was a staggering $25.5bn. That figure is rising again in 2008 – in tabloid terms to £1.7m an hour. Of course necessity is the mother of invention and cynics could argue that there is little alternative for Shell. Production of 'easy oil' reserves – those that are both close to markets and relatively simple to extract - are expected to peak by 2015. This deadline has challenged Shell to extend its search for 'unconventionals'.

Innovation Scorecard	
Innovation Culture	7
Strategic Innovation Focus	9
New Products	7
Managed Growth	9
Product Margin	9
Investment in Innovation	8
Innovation Brand Impact	8
Innovation Peer Review	8

Shell Key Data 2007	
Total Revenue	$355.7bn
Revenue Growth	11.6%
Income	$31.9bn
Income Growth	21.3%
Hyrdocarbon Production	3.3m boe / day
Total Assets	$267.5bn
# US Patents	144
Revenue / Employee	$3.4m

At the core of Shell's success in identifying new growth opportunities in recent years has been its Technology Futures initiatives that have been led by GameChanger – the company's corporate innovation programme. In 2003 the remit of the GameChanger team was shifted from an internal focus on idea exploitation to more of a strategic innovation role. Given the future challenges for the energy sector, Shell recognised the need to be more proactive in identifying and clarifying probable future areas of opportunity outside its traditional operations.

Shell already had a long-established reputation for long-term thinking with a dedicated scenarios team that produces a regular 20 year view on how the world is changing from a socio-economic and demographic perspective. What the new Technology Futures programme brought to the mix was an understanding of how technology from outside the energy sector fitted into the picture - in terms of how and where new technologies could both impact the world and, more specifically, the varied areas of Shell's current and future business operations. This programme led to the identification of a number of major potential spaces of opportunity as well as major threats to the business.

Amongst other outcomes, this programme led to significant investments in renewable energy technologies that will potentially develop into multi-billion dollar businesses. In one area, that of bio-fuels, Shell has been leapfrogging the mainstream. Where others are focused on grasping the immediate opportunity of transforming food into bio-diesel and bio-ethanol, Shell is interested in a more sustainable competitive approach. Driven by new insights, the company has been exploring next generation renewable feed-stocks such as agricultural waste as well as the production of marine based algae. Subsequent investments have secured future options and are building experience outside the company's traditional focus.

Within conventional fuel markets, in areas such as gas to liquid Shell has started to explore the viability of synthetic jet fuels as another way of improving emissions and fuel economy. Across the board, Shell's investment in innovation is paying off as it is gaining insight from developments outside its own sector and is finding opportunities in new areas. As the energy sector stands on the edge of dramatic change, Shell, one of its early pioneers, is well positioned to lead the move into a new generation of power options for the future.

SECTOR OVERVIEW

In January 2008, the price of oil product topped $100 a barrel for the first time. As this was almost five times more than it was in 2002, and with massively rising demand from the US, China and India taking place, the natural assumption was that this is due to the paucity of supply. This, however, is not necessarily the case. There is a lot of oil left, but most of this is not 'easy oil.' For one thing, the cost of extracting oil is increasing steadily. New oil discoveries tend to be found in inaccessible spots or in more unwieldy forms which adds to the cost. This doesn't mean that oil has run out. Canada's tarsands contain almost as much oil as Saudi Arabia. However it does mean that accessing these reserves in an economically and environentally acceptable manner is a far greater challenge than in the past. At the same time, many oil companies are finding it more and more difficult to find new reserves. In 2007, for the first time in 14 years, Exxon, the world's largest energy firm, failed to find as much new reserves as it had used during the year.

While wind, wave and solar attract much media attention, for now the economics of scale are not fully viable and these are more likely to be part of the longer term energy solution. In the meantime, as the pressure to migrate to clean or renewable energy sources is mounting, several countries from Finland to China and even the oil-rich UAE are making new investments in nuclear facilities, and new players such as ADM and Cargill have been pushing the bio-energy agenda at a global scale.

Across the sector, reducing CO2 emissions continues to influence policy. As the impact of climate change becomes more evident and general public awareness has increased, energy companies have been increasingly keen to display their green credentials and demonstrate they are actively pursuing and investing in renewable energy sources. The resulting financial incentives and markets for carbon will increasingly enable economic viable solutions around renewable energy production. That said, for now, the basics of the business are still largely dominated by the up-stream activities of oil and gas exploration and production and the down-stream activities of refining, distribution and marketing. Yes, there are new players in the sector – including a plethora of technology-based start-ups as well as the ex-state oil and gas companies – but the real power still resides with the 'majors' that continue to dominate the sector.

INNOVATION DRIVERS

Given the key challenges from rising global energy demand coupled with depleting major oil reserves in high-risk areas of the world, new oil and gas reserve discovery is a major capability that supports future growth. However, in the absence of any significant finds, creating new technological solutions to extract more from existing fields and accessing the less conveniently located reserves in deep sea, tar sands and possibly underneath the Antarctic is a major area of innovation opportunity. As off-shore oil production costs steadily increase, oil and gas companies are being driven to find clever ways to reduce platform personnel, operate multiple fields from a single platform and in some cases dispense with platforms all together by moving underwater.

Outside oil and gas, but still within traditional fossil fuels, there is also a growing focus on clean solutions to coal extraction and supply. This has huge impact for the likes of China and the increasingly domestically focused US. In the increasingly popular renewable energy arena, the major area for innovation activity is in creating economic solutions that can meet the long-term mass market demand and supply needs.

ONES WE ARE WATCHING

GE

With its ecomagination programme now in full sway, GE has been changing assumptions across much of its portfolio from jet engines through to medical systems. In the GE Energy business, it has demonstrated foresight and investment in new technologies across the space from wind and solar through to nuclear. Its photovoltaic systems are selling well, GE has installed over 8000 wind turbines and its ESBWR reactors are winning orders as the world gears up for new nuclear. At the same time, GE's recognised service business model expertise will certainly be having significant impact here as it has already done elsewhere.

BP

Although this previous Innovation Leader has had operational problems recently, its investments in innovation and new technologies have not abated. It has been busy cleaning up its oil and gas products, and, already a long time leader in the solar energy arena, BP has recently been making significant investments across the wider renewables area. Like Shell it is investing in next generation bio-fuels and has made interesting bets for the future in India and, most recently, with Synthetic Genomics where the potential for subsurface microbial processes are being explored.

Providing a differentiated range of branded offers in this highly commoditised and standardised sector

STARWOOD

Talk about a tax advantage! Starwood Hotels and Resorts was originally formed by the real estate investment firm Starwood Capital to take advantage of a tax break. Starwood is now one of the leading hotel and leisure companies in the world with approximately 925 properties in more than 95 countries. These cover a wide range from St. Regis, The Luxury Collection, Sheraton, Westin and Four Points by Sheraton, through to W, Le Méridien and the, recently launched, Aloft and element brands. It also owns Starwood Vacation Ownership, a premier operator of high quality vacation interval ownership resorts. Upmarket travel magazines often include Starwood properties among their top 20 hotels so this fast growing company is clearly well represented in all the best places around the world. It controls some of the most glamorous hotels around including the St. Regis in New York, Hotel Bristol in Vienna and Hotel Gritti Palace in Venice.

STARWOOD
HOTELS & RESORTS WORLDWIDE, INC.

Innovation Scorecard	
Innovation Culture	7
Strategic Innovation Focus	9
New Products	9
Managed Growth	9
Product Margin	7
Investment in Innovation	8
Innovation Brand Impact	7
Innovation Peer Review	9

Starwood Key Data 2007	
Total Revenue	$6.1bn
Revenue Growth	3.0%
Operating Income	$0.9bn
Operating Margin	13.9%
# Hotels	925
# New Hotels	29
# Hotel Brands	9
Revenue per Room	$160k

For several years now Starwood has had an edge on its competitors because of its strong brand awareness amongst its target markets and the way in which it has developed a range of differentiated experiences. The company's key asset is the associated diversification of the variety of brands, and the different market segments they serve, provide a broad base from which to enhance revenue. It aggressively attracts and cultivates new customers and maintains loyalty among the world's most active travellers: As one example, Starwood Preferred Guest, its award-winning frequent guest program, first made headlines when it launched with a breakthrough policy of no blackout dates and no capacity controls, allowing members to redeem free nights anytime, anywhere. Since then, it has grown to include more than 33m members and is cited for its hassle-free award redemption, outstanding customer service, dedicated member website and innovative promotions and benefits for elite members.

Although smaller than some of its peers, Starwood uses its scale to support its core marketing and reservation functions across its brands and reduce costs in areas such as insurance, energy, telecommunications, food and beverage, furniture, fixtures and operating supplies. Its size also limits exposure to any particular type of lodging, brand or geographic region. Strategically Starwood has also moved away from investment in owned real estate and increased its focus on its management and franchise business. In furtherance of this strategy, it has been selling some of the older hotels in its portfolio to, in turn, release capital to fund its ambitious international and branded growth plans. Simultaneously the company has retained one of the world's most profitable hotel net margins of around 15%, higher than many larger rivals.

Alongside these leading service based achievemets, Starwood has made a great success of its growing product innovation prowess. Having identified the bed as the key area for innovation opportunity in its mid-range Westin brand, Starwood designed the 'Heavenly Bed' providing the best night's sleep by far of any comparable brand. Not only did this well-targeted development improve the all important occupancy rates in Westin properties, it also created a new business line as customers started to buy the Heavenly Bed for their homes. This has now been extended into the W chain where whotelsthestore.com now enables anyone to buy the key ingredients of the unique W experience. From the W bed, pillows and towels to books and an increasing range of male and female apparel, this is the latest incarnation of high-impact product innovation from this service sector pioneer.

SECTOR OVERVIEW

Luxury hotels are enjoying a come back. A survey of 2007 hotel prices worldwide found that five star occupancy rates were up 15.2% compared to growth of just 4.8% in the budget sector. More people are travelling and they have put away their backpacks to do it in style. This has been helped by relatively continuous economic growth and the prevalence of low cost airlines which have enabled consumers to take more trips on a regular basis. Despite a dip in consumer spending, luxury hotels will remain big business and as long as people continue to revel in the experience escape, the outlook appears positive. The shift in focus of luxury moving from 'product' to 'experience' is indicative of the evolution of customer experience and expectation and is an area where several brands are making a mark.

Luxury hotel markets have been booming for the last few years and there is huge investment interest in this segment. From Dubai to Shangai and New York to Paris, this trend is coming from consumers, investors, developers and owners alike, who see hotels as undervalued assets with good collateral in an expanding market. In the first half of 2007 in the US alone, more than $11.5bn changed hands in upscale and luxury lodging. The sector has been particularly active due to recent increase in demand, short supply, and lead times for construction. This, coupled with an anticipated continued increase in revenue per available room, has created an enticing environment for investors as evidenced by the takeover of the merged Hilton Hotels Corporation by private equity group Blackstone for $26bn during the summer of 2007.

INNOVATION DRIVERS

It's luxury with a conscience. Ethical holiday directory, responsibletravel.com announced in February 2008 that it had seen an 87% increase on the number of people enquiring about responsible vacations. These include helping local people to earn a fairer income from tourism, supporting local conservation or ensuring that any negative environmental or cultural impacts are reduced. Such environmental awareness is more common among independently operated properties than international chain hotels, but with new government incentives in place, running a green hotel is becoming more economically feasible. Public and media pressure has meant that several hotel brands are now boosting their eco-friendly credentials and brands within established groups are differentiating themselves by addressing the issue. For example Aloft by Starwood Hotels is using recycled teak in its properties and has provided special parking spaces for hybrid cars; Hyatt's Andaz sub-brand promises to incorporate eco-friendly materials into all new builds; Fairmont's Green Partnership focuses on waste management, energy and water conservation and community outreach programmes; and Hilton hotels have launched the 'we care!' programme across 79 properties in Europe and Africa.

Co-branding is also on the up. Many luxury hotel companies are pursuing out-of-room revenue growth opportunities, including signature restaurants as well as spa and associated product innovations in specific properties. This has led to, amongts others, Gordon Ramsey at Claridges, Armani and Bulgari fashion house hotels and the association with other 'cultural brands' such as Bliss spa treatments. For an increasingly picky clientele the hotel experience is no longer just about room size, bed quality and location. While some may see this high-end focus as a niche, the trickle down effect across key mass brands is having significant impact. The core chains from Hilton and Hyatt through to Sheraton and Marriott are variously seeking to capitalise on the association with the Park Hyatt, St Regis, Ritz Carlton and similar up-market sub-brands in their portfolios. For the majority of hotels the core Monday to Thursday crowd is the business community for whom a balance of cost, convenience and comfort is the crucial influence. As the area where Value Innovation first had significant impact in the late 1990 providing the service combination that meets the critical business travel criteria is still a major area of opportunity.

ONES WE ARE WATCHING

Hyatt

Hyatt gained fame with the opening of the world's first atrium hotel in Atlanta 1967. Now one of the largest hotel chains around its portfolio spans a wide range from value suites through to premium properties: Alongside the new ANdaZ brand that launched with the Great Eastern Hotel in London, at the top end of the market the Park Hyatt chain is focused on personalised experience with a boutique feel. The Park Hyatt Toyko, its flagship property, was featured in *Lost in Translation* and the company has been extending the brand, most notably to Chicago.

Jumeirah

This Dubai owned group has been investing heavily in premium properties around the world. As it seeks to become a major player in the industry its acquisition of such prestigious hotels as the Carlton Tower in London and Essex House in New York are complementing its collection of top locations back home. As Dubai, and the UAE as a whole, promote massive growth in the number of visitors coming either for stop-overs or medical tourism, Jumeirah is showing signs of leadership in the next generation of tailored hotel experiences.

Pragmatic, customer focused innovation, driven by tangible insights and delivered with class-leading efficiency

TESCO

Tesco is Britain's leading food retailer and the third largest in the world. Its first store was opened in 1929 in London and by the early 1960s Tesco was a familiar feature of most UK high streets. After joining the eighties trend for large out-of-town supermarkets, in the 1990s the company started pioneering many new innovations. It developed new store concepts such as Tesco Metro, a city centre store meeting the needs of local shoppers, and Tesco Express, the first UK petrol station convenience store. In 1995 the company introduced its Clubcard, the UK's first customer loyalty card, and two years later formed a joint venture with the Royal Bank of Scotland to offer a range of financial services. 2000 marked the start of Tesco.com which was built on the back of existing stores and, with low capital spend, was profitable from the start – a key internal requirement. Tesco's international operation, which started in 1994, has steadily expanded and now accounts for half of its total retail space. Since 2000 there has also been an increasing focus on building non-food sales both in store and online with the result that, for example, Tesco is now the UK's largest CD retailer.

TESCO

Innovation Scorecard	
Innovation Culture	8
Strategic Innovation Focus	8
New Products	9
Managed Growth	9
Product Margin	7
Investment in Innovation	8
Innovation Brand Impact	7
Innovation Peer Review	9

Tesco Key Data 2007	
Total Revenue	£46.6bn
Revenue Growth	10.9%
Pre-tax Profit	£2.7bn
Profit Growth	20.3%
# Stores	3263
# New Stores	591
UK Revenue / Sq Ft	£1,325
UK Revenue / Employee	£177k

Innovation in Tesco is seen as key for its customers and also to its business. As such the company focuses equally on product as well as process innovation. Core to Tesco's innovation success has been its focus on managing margins. Right from the start the company's obsession with efficiency has been used to keep prices low and, in 2007, Tesco managed 1.8% deflation across its product lines. The company has also focused on making the shopping experience as easy as possible for customers – be it in hypermarkets, small stores or online. Tesco Express has led the move of food retailing back into the community and Tesco.com now generates sales of more than £700 million and profits of more than £35 million for the business. In the UK, the service reaches 96% of the population and handles more than 170,000 orders a week.

Like some of its peers, Tesco also aims to improve service and provide better value rather than concentrate on pricing alone. These principles are carried across the business into non-food, services and its international operations. To enable this, the company pays considerable focus on harnessing the creativity of its workforce and encourages staff to come forward with ideas. The company's prowess in process management applies just as much to its idea management as it does to logistics and store layout.

A key ingredient to Tesco's growth is the use of well-targeted own-label brands including the up-market 'Finest' and low-price 'Value' labels. To drive this Tesco has led the field in market insight. Its Clubcard, the most successful loyalty card in the sector, provides Tesco with a class-leading ability to spot emerging trends, attract consumers and influence the behaviour of secondary customers to bring them into the fold.

Another influence has been a coherent strategy for international expansion. As well as expanding into emerging markets in S.E. Asia and Eastern Europe, Tesco is now also tackling the US where its first Fresh & Easy stores opened in 2007. The company's innovative format differentiates it from existing, homogenous American supermarkets, whilst capitalising on a gap in west-coast grocery retailing. Tesco's expansion of its Fresh & Easy business is hard to go unnoticed. The promise to deliver its 'Every Day Low Price Plus' strategy is set to ensure that Fresh & Easy is well placed to capitalise on US customer's search for quality and value, especially as consumer spending decreases. As it drives growth in new territories and dominates its domestic market, Tesco sees that innovation success is down to four issues - trying new things; not being complacent; being prepared to change; and being determined that no other business will achieve more for customers.

SECTOR OVERVIEW

The supermarket and the one-stop-shop transformed the dynamics of the food and general retail sectors. Pioneered by the likes of Sainsbury's, Safeway, Carrefour and Wal-Mart, the advantages are clear; securing large volume discounts allows supermarkets to undercut traditional stores and greater space means they can offer as good if not better selection of produce. In recent years, there has been a focus on multi-format capability and blurring of formats as retailers 'mix and match' to increase effectiveness and offer customers a reason to shop at their stores. Linked to issues such as out-of-town planning constraints this has led to consolidation including the opening of smaller stores within communities, the incorporation of pharmacies, post offices and travel agencies, joint retail ventures with the likes of BP, Shell and Exxon and the growth of on-line shopping. In addition, the rise of own-label products and an increasing migration to non-food have fundamentally changed the product mix. By 2010, the value of the European private label market is forecast to reach €430.8 billion, up from €298.1 billion in 2005.

With price deflation and commoditisation a threat in most markets, genuine innovation remains the key for many retailers. In more mature markets there is a growing appreciation that a 'one size fits all' approach is no longer valid. Some retailers are adapting by focusing on other differentiation than just price using new channels to develop tailored products and services to suit more sophisticated customer requirements. Growing consumer spending and a hunger for modern retail in emerging markets such as China, India and Russia also provide significant opportunities for growth. Large international and regional grocers are acquiring market share across the world at a remarkable rate with the top 100 retailers already capturing 45% of the world's modern grocery distribution. Companies like Wal-Mart, Ahold, Carrefour and Tesco now face the challenge of managing increasingly complex supply chains, building critical volume in different countries and maintaining comparable levels of product quality and customer experience.

INNOVATION DRIVERS

In this highly competitive arena, innovation is rife. From introducing new technology and broadening the product portfolio to positioning the brand and deepening customer relationships, there are several key drivers gaining widespread attention. Today's two main technological sources of innovation are the on-line provisions market and an improved supply chain-focused technology such as smart tags. Both are addressing improved efficiency of goods supply and provision, but are also areas of consumer-focused innovation around convenience and traceability. In terms of broadening the product mix, the migration of food to general product supply has been followed by service development using the retailer brand as the focal point. Starting with a limited range of joint venture financial products, this has spread to loans, insurance, holidays, car retail and, most recently, energy supply, virtual mobile phone operations and real estate. All are usually delivered in partnership with leading existing suppliers, but are positioned around the increasing levels of trust that consumers have with their favorite retail brand.

Many retailers are using the migration of their loyalty cards into their joint venture services and relationships such as Nectar in the UK to gain new information about their wider customer base that feeds into detailed segmentation analysis and opportunities for cross selling. Retailers can now find out what their customers spend their money on outside the store and can use this to develop new branded service propositions and improve the levels of cross-selling between groups. Together these are all providing the leading retailers with the opportunity to take consumer-centric innovation to a whole new level. Lastly, as efficiency is such a driver of margins in this sector, internal process innovation is a correspondingly key area of focus. Customers largely see the visible impacts from innovation in terms of new products, services, store layouts and website usability. But as new technologies hit the supply chain it is the companies that can best manage the flow of materials that will win the day. As such an efficient IT infrastructure is vital and this has been a major area of spend in recent years for the main players.

ONES WE ARE WATCHING

Wal-Mart

The world's largest retailer has been through a number of challenges over recent years but is now making a major success of its innovation focus on the sustainability agenda: Wal-Mart has an ambitious strategy to reinvent itself as a champion of the environment. The company plans to eliminate 30% of the energy used in stores, reduce solid waste from its U.S. stores by 25% within three years, and invest up to $500 million in sustainability projects. Wal-Mart is creating competition in its supply base around being green that is already having significant impact on innovation in packaging and waste management.

WuMart

At the other end of the scale at the moment, China's WuMart is seen by some as the future of retail. The company has only been around since 1994, and although only having $320m of sales and 500 stores at the moment, this is one of the fastest growing retailers in China with grand expansion plans and a different take on customer service. The company tailors stores to suit local tastes and fresh food is a major feature, aware, for example, shoppers can pick live fish to have for dinner. With 700% growth predicted for retailing in China alone over the next decade, WuMart is intent of having a major role in this.

Clearly prioritised innovation across the board delivering focused impact into the global marketplace

UPS

United Parcel Service (UPS) celebrated its 100th anniversary in 2007 and is the world's largest package delivery company and a leader in supply chain services. Offering an extensive range of options for managing the movement of goods, information and funds, it is headquartered in Atlanta, serves more than 200 territories, is the 9th largest airline in the world with nearly 300 aircraft and has just under 100,000 vehicles in its delivery fleet. It has been quick to take advantage of emerging opportunities over the years. Its first air-express flights started in 1929 and an intercontinental service between the US and Europe was launched in 1975. The UPS website has proved to be a vital platform for growth and now provides 15 million customers a day with on-line tracking. While its primary business is the time-definite delivery of packages and documents, UPS has also extended its capabilities to encompass a broader spectrum of services.

Innovation Scorecard

Innovation Scorecard	
Innovation Culture	7
Strategic Innovation Focus	9
New Products	7
Managed Growth	7
Product Margin	8
Investment in Innovation	8
Innovation Brand Impact	9
Innovation Peer Review	9

UPS Key Data 2007

UPS Key Data 2007	
Total Revenue	$49.7bn
Revenue Growth	4.5%
Total Operating Profit	$578m
Delivery Fleet	94,542
Packages Delivered Daily	15.8m
# UPS Stores	4,800
Employees	425,000
Brand Value	$12.4bn

UPS is a leading provider of supply chain solutions, such as freight forwarding, customs brokerage, fulfillment, returns, financial transaction and less-than-truckload transportation services. Hence its vision now encompasses a broader view of logistics – 'to synchronize the world of commerce, managing the complexities of our customers' supply chain needs'.

Central to its recent growth has been an innovation model that is used for many strategic decisions. This differentiates new business and product concepts along two dimensions – internally versus externally focused opportunities and innovations to be driven through existing businesses versus entrepreneurial ventures. Each area within the model has different criteria, different approaches for delivery and different organisational and financial requirements that guide a wide range of innovation activities. On one side, these include internal product development focused on new services for customers, where UPS was recognised as a leader by *Business Week*, as well as other 'can't fail' external opportunities that impact operations. On the other side, UPS has been pursuing a 'can fail fast' strategy with more entrepreneurial innovations and also established a strategic enterprise fund that invests in such opportunities as tagging and fuel-cell technologies.

A recent example of UPS innovation is a new service - UPS Paperless Invoice and International UPS Returns which was launched in January 2008. UPS are the first package carrier to offer customers a paperless international shipping option as well as a package return capability. The majority of the small packages UPS ships internationally - everything besides a letter or document - require a commercial invoice. UPS Paperless Invoice allows shipments to clear customs using electronic data in lieu of these error-prone forms. Moreover, because information is stored electronically, customers reduce their chance of making mistakes when filling out customs documentation.

Other enhancements have included the expansion of Worldport, the main air hub in Louisville, Kentucky, which now has a European counterpart in Cologne, and another is being built in Shanghai, making UPS the first US airline to open an international air hub in China. Asia has been a primary growth target for some time and, over the past five years, UPS has invested about $600 million in China and, in 2007, also announced significant expandion of its services in India. As UPS delivers 15.8m items a day and applies a rigourous approach to innovation-driven growth, its position, its leadership is strong and well-supported.

SECTOR OVERVIEW

Essentially covering the whole process of planning, implementing and controlling the flow and storage of goods and services from the point of origin to the point of consumption, the logistics industry has been in a state of continual growth throughout the last century. Now covering capabilities such as purchasing, supplier management, materials handling, inventory management, warehousing, distribution and transport, all with increasing levels of customer service, it has migrated from a plethora of independent operators into integrated global entities and alliances. With a growing demand for 'just in time' manufacturing and distribution systems throughout the 1980s and 1990s, companies in the supply chains which feed end-product manufacturers were rated not only on the quality of their goods but also on the speed and accuracy of delivery. Effective internal logistics capabilities became a key source of competitive advantage and the strategic focus for companies as diverse as Toyota and McDonalds.

As many industrial and retail firms have focused on their core businesses, third party logistics, where an outside organisation is used to manage all operations from information systems to customer order processing, has been a particular area of growth. Federal Express provided a major push for the logistics sector with what it sees as the invention of an integrated air/ground network. Essentially this and similar moves towards internationally integrated global operations taking care of the full door-to-door process have given rise to developments such as the expansion of national postal services and a series of acquisitions by the likes of FedEx, UPS and DHL to broaden their coverage and range of services. Off-shoring and e-commerce have also had a major influence; the first in increasing the volume and frequency of international package delivery and the second, particularly with sites like eBay, by increasing volume and the need for confidence and trust in successful completion of delivery.

INNOVATION DRIVERS

Technology continues to be the main source of innovation in this sector. After the recent growth of some of the main players to accommodate an increasingly large customer base, the focus is on improving clarity and reliability. In the move towards the fully integrated and transparent supply chain, greater visibility of where individual components and packages are as they whiz around the world is a major customer need and so is an opportunity for logistics companies to add extra value. Web-based package tracking was a major move forward for the customer and internal innovation is further improving the speed and accuracy of package registration and monitoring. A key technology in this last area is RFID, which, through its ability to have unique information remotely read and written, is transforming the way in which the whole sector functions. Ultimately on track to replace barcodes globally, this technology has been one which has attracted much attention, both negative in terms of its role in enabling a potential big-brother society where everything from products to people can be tracked, and positive in terms of the associated improvements in supply chain efficiency.

According to McKinsey, global package and freight shipments are expected to more than double over the next decade with an estimated 85 percent of manufactured goods sold across borders by 2020, compared to 20 percent in 2002. As global infrastructures become more sophisticated and mobile computing becomes pervasive in such markets as China and India, the essential building blocks for even more efficient logistics operations are all coming together. Cheaper RFID technology, more intelligent software for both dock and yard management and more transparent web-based tracking are all helping delivery to be more secure and timely and inventories to be minimised. Allied to ever increasing volumes of freight, this ever more accessible technology is going to place more and more emphasis on the leading logistics firms to innovate around their business models as much as the products and services that they provide to their global customers.

ONES WE ARE WATCHING

FedEx

With the acquisition of Kinko's now paying dividends, FedEx is deepening its opportunities for wider consumer interaction with an increasing range of global services. As well as implementing new development processes, the company has set up an Innovation Lab to explore technologies such as mobile tracking, biometrics and video object recognition that will have impact in the medium and long term future.

TNT

Dutch-based TNT has refocused and is growing steadily. TNT Express has been improving the customer experience in its home European markets and has also launched the industry's fastest express delivery service in South East Asia. One major area of innovation focus across the company has been in and around the sustainability agenda – an area where TNT is aiming for sector leadership.

Redefining the consumer experience across all levels of the product, service, innovation travel portfolio

VIRGIN ATLANTIC

Virgin Atlantic has been redefining the airline experience since the day it was launched in 1984. As part of the Virgin empire, it has, not surprisingly, been challenging the establishment, championing the consumer, improving service and astounding its customers while simultaneously making money. Having repeatedly won best airline awards throughout the years, it has been consistently using innovation as a core part of its strategic approach. Often imitated, but so far never usurped, Virgin Atlantic is the UK's second largest airline, has won numerous accolades for its use of design, its service and its promotion. Above all, over the years, Virgin Atlantic has been a pioneer of customer visible innovation. Its long list of firsts include being the first airline to have individual TVs in all classes, to provide child safety seats, to offer a super economy service, to have no smoking flights, to have drive-thru check-in, to have an arrivals lounge, to allow mobile connectivity and SMS texting on board, the first to have at-seat podcasting and, most recently, the first to fly one of its planes using biofuels. Although not the largest airline around, being first in all other aspects seems to have become a trademark of the Virgin Atlantic approach.

Innovation Scorecard	
Innovation Culture	8
Strategic Innovation Focus	9
New Products	8
Managed Growth	9
Product Margin	8
Investment in Innovation	7
Innovation Brand Impact	10
Innovation Peer Review	10

Virgin Key Data 2007	
Turnover	£2.14bn
Profit	£47m
Destinations	30
Aircraft	38
Average Age of Fleet	5.3 years
Passengers Carried	5.9m
Market Share (London – NY)	20.1%
Employees	9877

Having changed the perception of business class travel with the introduction of its Upper Class service, which provided 'first class experience at a business class price', Virgin Atlantic soon caught the target market's imagination. As competitors responded with similar products, Virgin Atlantic then followed up with the introduction of the Upper Class Suite in 2003. This multi award winning concept introduced a flip over chair providing a completely flat bed and a wider seat than any alternative airline. In addition, it also included an in-flight cocktail bar, on-board massage as well as personal limo services for customers.

Moving on from on-board innovation, in 2006 Virgin Atlantic opened its new £11m Heathrow Clubhouse that fundamentally changed the perception of what a lounge could be. More like a cross between and night club and a private club, it contains a myriad of features ranging from a long bar, a deli, a brasserie, a poolside lounge, a library and a rooftop garden through to a Cowshed branded spa which houses four massage bays, a sauna, six steam-shower rooms, sun-tan booths and, for those fancying a quick dip before flying, a hydro-pool! While this is all very attractive at a functional level, from a design perspective the implementation is exemplary.

And of course all this is 'free' to Upper Class customers. It might seem like a lot of marketing showmanship, but there is solid business logic behind it all. Press coverage, multiple industry awards and word of mouth migration of customers to Upper Class, either from rival airlines or upgrading from economy, has been substantial. Most recently the company upgraded its Premium Economy seats and added to the Upper Class experience with a 10 minute check-in from limo to lounge which, at Heathrow, is almost a miracle.

The key to all this has been a world leading design team that focuses equally on product innovation and service innovation. Charged with profitably redefining the consumer experience, this team is well known for changing the rules and giving customers new innovations and experiences that most of them would never think of. Virgin Atlantic is the true sustained innovation leader of this sector. The company is determined to continue to grow by continuing to deliver incessant high-impact innovation and, despite increasing competition from a select few Asian and Gulf-based operators, Virgin Atlantic is showing no sign of handing over its leadership position any time soon.

SECTOR OVERVIEW

The airline industry has been through a lot of turbulence over the past few years. In the wake of 9/11 passengers stopped flying, most airlines started making huge losses and several went into Chapter 11 protection. Especially in the US, the airline sector became a byword for poor financial returns. As price discounting and increased security helped to build passenger numbers, a new range of leaders emerged in the sector. The old guard of primarily national flag carriers such as BA, Air France and American were quickly eclipsed in terms of financial performance by a raft of new low-cost airlines that soon came to dominate short-haul routes. SouthWest Airlines in the US, EasyJet and RyanAir in Europe and AirAsia in Malaysia are just some of the stars of the low-cost flight fraternity that have changed the way many of us now fly. Through opening up air travel to previously under-served consumers, millions now pay rock-bottom prices to get basic service as they are often flown to small, remote regional airports. It may not be the same service as you would get from the established airlines on the traditional routes, but with discounts of up to 90% on scheduled prices nobody has been complaining. With RyanAir consistently worth more than BA, low-cost airlines are now firmly established as part of the airline sector. New ones come and go but their presence is now nearly global and dominates much of the short haul routes.

This has meant that the other airlines have increasingly focused on the more profitable long-haul routes flying across the Atlantic, Pacific or Asia. JFK to LHR is the busiest and most profitable trans-Atlantic route but with the open skies agreement now coming into force, more airlines are free to fly to the key destinations. At the higher end of corporate travel, fractional jet ownership from the likes of Net-Jets is now delivering major point-to-point disruption of a market which is now also being targeted by business class only transatlantic services from the likes of SilverJet. Across all these varied flight options, the key dynamics by which the sector measures performance however remain constant - passenger volumes, yields (average fares) and load factors (seat occupancy) - all of which are oriented around maximising asset utilisation and margin growth.

INNOVATION DRIVERS

An ever present problem with the airline business is that the core product, the flight from A to B is largely a commodity. All airlines fly similar planes that take the same time to cover the distance and tickets also cost roughly the same across economy and business class. Such price matching between most airlines implies the need to attract target business class customers in different ways. Moving people from economy to business class and premium economy is the name of the game. Within an increasingly competitive and security conscious market and with increased fuel prices, margins are cut to the bone. In the absence of any really major technology breakthroughs, the main drivers of consumer-focused innovation for major airlines continue to be service, price, punctuality and alliance membership. Singapore Airlines, RyanAir, British Airways and the Star Alliance have respectively been strong in these areas. The key challenges for established national carriers and the largely low-cost and business class focused start-ups alike have been to create and deliver sustainable, profitable combinations that attract and retain one or more of the three primary customer groups – business travellers, city-to-city economy and mass-market / low-cost. As the volume of those flying in each of these customer groups continues to rise steadily, innovation around service has been gaining increasing priority. Comfort and entertainment as well as the quality, or lack of, in-flight food have all become priorities as airlines strive to maximise volume and load factors. Another innovation driver in the sector is reducing the much discussed carbon footprint. While some airlines have started to offset their emissions, few have yet to make a break-though. The exception to this seems to again be Virgin Atlantic which in 2008, in partnership with GE and Boeing, was the first airline to fly a plane running partially on bio-fuels. And its founder, Richard Branson, is donating all his profits over the next 10 years into renewable technologies research and has also set up a $25m prize for the first person who can remove 1billion tonnes of greenhouse gases from the atmosphere.

ONES WE ARE WATCHING

Qatar Airways

Qatar Airways is on the fast track with an average of 35% growth year-on-year for the past 10 years. It is looking to quickly capture a significant share of the long-haul market, make Doha a key global hub and attract a good proportion of the first and business class market while it is at it. As such, exceeding the experience available with other airlines is critical and Qatar Airways has not held back with a unique First Class lounge for passengers to relax and unwind at 35,000 feet. As one of only five airlines to have been awarded a five-star rating by Skytrax and now with the world's first passenger building dedicated to First and Business Class passengers, this airline has #1 position in its sights.

Singapore Airlines

Asia's leading airline is consistently voted best-in-class for cabin service and is now taking innovation to heart. Now, as the first company to fly the Airbus 380, Singapore Airlines is busy upgrading all its products from economy up: It has introduced Singapore Airlines Suites, a brand new class beyond first, offering the ultimate in luxury travel in your very own personal cabin including sliding doors, window blinds and a standalone bed. With a 49% share in Virgin Atlantic, Singapore Airlines has been picking up tips as it rolls out its new services. With strong competition from new regional low-cost flyers, it has also invested in 49% of Tiger Airways which is already expanding into Australia and South Korea.

THE COMMON INGREDIENTS

At a high level, across the twenty five companies we have profiled as the current innovation leaders in part one, there is little doubt that innovation is being delivered with major impact. As highlighted in the varied analysis, new products, services and processes are successfully pushing the boundaries, exceeding customer expectations, capitalising on new technologies and generating higher revenues and increased margins. Across the firms profiled here, five common traits are evident, and these can be seen as the common ingredients for innovation competence today

Strategic Focus – There is a strong strategic focus on the role of innovation within their markets and the contribution that innovation makes to the business. All of these companies know where innovation is going to make an impact and where it is not. As such they focus their innovation activities in the most appropriate areas, whether that be primarily on new business opportunities like Shell, on the deepening the customer relationship, as is the case with Tesco and Handlesbanken, or on disrupting adjacent sectors, as demonstrated by Google.

Insight – They have an excellent understanding of the marketplace, their customers and an ability to configure products and services around emerging needs. Most firms have access to masses of market research, consumer insight, emerging trend data and new technology foresight but few actually use it. The innovation leaders profiled here all, without exception, access, create, interpret and exploit the insight they need to provide both the context and focus for new innovations.

Collaboration – They clearly understand both their core capabilities and those of their partners with who they work together to deliver innovative products and services. Aviva is one obvious case in point here as the company openly partnered with IBM and Orange but the same approach also can be in other leaders. Virgin Atlantic collaborates with the likes of Soho House in developing new service innovation approaches; Nokia and Google are working together across many areas; and H&M has been making collaboration with designers and celebrities a unique platform for innovation.

Process – They have simple yet effective approaches to conceiving, qualifying, developing, and then quickly launching, new products and services. Reckitt Benckiser is clearly exemplary in the way it identifies an unmet consumer need and rapidly launches new products into the market without excessive bureaucracy, but Apple, Nokia and Samsung all have and use the same capability. In addition, in the services arena, the likes of Tesco, Starwood and UPS all have fast and efficient development processes that enable innovation to be delivered in a timely manner.

Organisation – Roles, responsibilities and culture all support innovation while appropriate metrics are used to measure and reward successful innovation. In some companies such as Shell, Starwood and Virgin Atlantic there are clear leaders of innovation activities – be they called Global Innovation Directors, Chief Innovation Officers or an equivalent. In others, innovation is more widespread across the organisation. Apple, Adidas, Google, Nokia, NTT, Reckitt Benckiser and Tesco would all see that innovation is more pervasive and embedded into the wider corporate culture.

These five traits are not unique to these firms – some would even see them as 'hygiene factors'. However, in all cases, they are certainly present, they are well integrated and leveraged to best effect. Five years ago, being good at one of these five would have been enough for some firms to gain competitive advantage in their sector. Today this is definitely not the case.

These five traits are not unique to these firms – some would even see them as 'hygiene factors'. However, in all cases, they are certainly present, they are well integrated and leveraged to best effect. Five years ago, being good at one of these five would have been enough for some firms to gain competitive advantage in their sector. Today this is definitely not the case. With the advances that have been made in the way companies focus, manage and deliver innovation, any company seeking to even be on the same playing field as the companies profiled have to be exemplary at all these core common elements.

Alongside these, five main traits, the Innovation Leaders of today are making a difference by also pioneering the adoption and exploitation of a number of other approaches. Some of these, such as customer-centred innovation, Open Innovation and identifying future growth, are well established in some fields and are now rapidly migrating across to other sectors. Others, such as smart M&A and multiple innovation, are more emergent and are only just being recognised as having widespread application. In the following chapters we explore these and other issues in more detail.

These new approaches are being used by the varied Innovation Leaders to complement excellence in the five common ingredients identified above. They are not replacing the core internal capabilities upon which their innovation prowess has been developed. Some companies, keen on gaining a jump-start on others, may well be tempted by the examples that follow and seek to race ahead with using these approaches. For most this would be an error. The new approaches work best when built on top of, and not instead of, the five core capabilities.

PART 2 - ADDITIONAL INSIGHTS

CUSTOMER CENTRED INNOVATION

Bringing the customer more and more into the heart of the innovation process is a critical issue for many firms. Although initially only a major concern in the more consumer-focused sectors such as sportswear, fashion retail and consumer electronics, being more customer centred has now become a common feature across the likes of BASF, Handelsbanken, NTT DoCoMo and UPS. As such most companies now seek to gain impact by increasing focus in this area. Actually, user-centred or, as some have recently sought to rename it, people-centred design, has been on the academic agenda for over twenty years and has been practiced in one format or another since the early days of Xerox PARC. Whether focused on human-computer interface design, physical product use or finding the right positioning of a service in the consumer's mind, the broader field of customer centred innovation is now a key part of many organisations' development and growth activities: To innovate without a customer centred focus has, for many, become an anathema. As this area gets ever more crowded, many are asking both for better clarity on the relative merits of the different approaches being touted and also for greater understanding of which way this increasingly significant market is itself moving.

At their heart, all of the varied approaches to customer-centred innovation that are being used today are seeking to gain greater focus around what the true needs of target users are and how can they best be met, or even exceeded, through the product or service being developed. To support this, a number of different philosophies have been introduced over the past few years to help consumer research and product development communities to work together better.

These have included both 'Voice of the Customer' which became a general term used in the product development community in the 1980s to highlight several approaches for bringing the customer perspective into the innovation process, and 'Six Sigma' - a more specific approach that was originally developed in Motorola in the late 1980s which later grew to encompass a customer-centric view for new product development and delivery.

As companies have sought to better gather relevant consumer research and apply the associated insights in the most appropriate manner, there has been a shift in their use from the concept validation phase of a development to earlier stages such as core strategic insight acquisition and interrogation - right at the start of the innovation process. While there are numerous flavours of approaches for increased customer focus being promoted by varied interested parties, at a generic level there are four main fields of activity:

1. Personas
2. Ethnography
3. Fan bases and
4. Participatory design

Each of these has areas of specific application that have been, and continue to be, exploited as a key part of the innovation process by a number of leading organisations.

Personas

The most established user-centred design approach of creating personas is in widespread use. These are fictional characters, often living in the future, whose needs, behaviours and attitudes are based on combinations of interviews, research and customer profiling. They are not linked to one specific product or service but, rather, can be used as a test of relevance for new products. In some organisations up to 20 personas are used to provide a rich and varied portfolio of potential users for referencing and stimulation. They enable a range of social, emotional and cultural needs to be integrated, shared across a development team and used as stimulus for innovation:

In Philips, where personas have become a core part of the organisation's 'people research' and concept development approaches, they are used to provide context

for innovation. With a rich repertoire of ongoing social, consumer and technology trends research taking place, a key challenge was how to best capture the insights and feed then into the design teams developing new product and services concepts across the company's wide spectrum of activities that range from healthcare and lighting through to consumer electronics. Using standardised templates that include images, text and data, personas are created in a number of formats – sometimes as diaries, sometimes as interviews and sometimes as articles.

If undertaken effectively - by which Philips sees that they must be believable, tell a real story, describe a person and their needs in context, combine segment characteristics and convey a clear character – personas act as vivid bridge between the customer and the development team: Projects can either be driven by and linked around one or more specific personas for focus, or a wide range of personas can be used to test applicability of new concepts.

In Microsoft, personas have been used to shift a technology-centred culture to one that is more customer-centric. They have helped to develop recognition of people's behaviours within social, emotional and cultural contexts. Alongside giving focus for their software and hardware designers alike, personas are seen to 'provide a shared basis for communication' especially around understanding who the intended users for a new product release really are. Now employing over 800 product designers and 150 user researchers responsible for evolving the usability and interaction of its products, Microsoft continues to use personas as a key vehicle in bringing customer needs to life for its integrated product development teams.

Within Intel, the chip company, personas have also been used to drive increased understanding of both existing and emerging global cultural experiences. Intel's Innovation Strategy Team use personas created by the firm's People and Practices Research group to uncover new potential computing applications, shift internal preconceptions within the business and challenge the company's business model – especially around future technology development. Again this helps to both discover and guide new innovation.

Ethnography

Sometimes dubbed 'industrial anthropology', ethnography has been getting a lot of attention in the business press – from *Business Week* and CNN to the *FT*. It focuses on the observation of target users in their environments of use – be they home, work, leisure or retail – and seeks to identify key needs and behaviours that the individuals don't necessarily talk about in interviews but do exhibit in practice. Originally developed from its anthropological routes into a core research activity for innovation and product development at Xerox PARC, ethnography has become one of the core consumer-insight tools: By the late '90s nearly all significant design groups had ethnographic capabilities within their teams.

In Microsoft, detailed ethnographic research is undertaken to directly support the development and updating of the personas outlined above. In addition initiatives such as 'adopt a family' where developers became personal IT consultants to families are used to provide developers with new insights on the in-use usability of their software.

Intel is interested in influencing the ecosystem around its products and their end-use and, as such, is keen to understand how people's behaviour will change on a 5 to 10 year timescale. Intel's People and Practices

Research team seeks to identify future opportunities ahead of its direct customers. They do this so that they can anticipate the associated emergent needs, uncover new product ideas and pre-empt their potential impact and application both with the PC market and in attractive high-growth adjacent sectors such as mobile communications. Recent projects have included 'The Next Ten Percent' which used ethnographic fieldwork in Asia and Latin America to identify new users and usage models and highlight the differences between these and standard US / European approaches. As with many other organisations, Intel is focused on how the wider market is changing a period of time out from today that roughly equates to twice the company's natural innovation clock-speed. As such, both its internal ethnography teams and the Intel Research 'Labettes' it has established with major research universities often take the ten year view.

eBay sees itself as being more 'user-driven' than 'user-centred' and its development projects tend to focus on the value that customers will associate with new services. Ethnographic research looking between 2 and 7 years out is linked directly into shorter term projects which are steering user-interface developments on a 6 to 12 month schedule. Often incremental in nature, the associated innovations then gain further consumer input via customer forums, message boards and customer conferences. Addressing the future needs of both current and target customers, the people focus for eBay is very much driven by the financial upside that can be delivered.

Fan Bases

Fan bases are groups of 'lead users' who are engaged directly within the development of new product and services concepts. Through a variety of means ranging from consultative groups, peer reviewers and on-line dialogue, they are engaged by the internal innovation teams to help steer concept development from the earliest possible stages. To feed into its ethnographic and user interface projects, eBay has around 1000 global users who are engaged on a weekly basis by a dedicated user experience team. Similarly, Microsoft has a web-based volunteer panel of 6000 product users from whom feedback data is collated on a regular basis and fed into the development teams. While other organisations from WHSmith and Staples in the retail sector through to GE in the financial services space also use fan bases actively to solicit external opinion for new product development, two of the most significant users of this approach at the moment are BMW and LEGO:

BMW is a strong advocate of what it terms luminaries – lead users recruited to help steer new vehicle feature development. While the company has a number of activities such as technology scouting introducing specific 'technologies of tomorrow' that will impact the driving experience, from a marketing / customer perspective, the company looks for general external marketing inputs from a wealth of leading internal and external suppliers supported by specific customer insights. These insights are gained through a combination of participative concept hot-house testing within customer innovation labs, pre-launch car clinics and innovation road-

Several Companies have started to make greater use of lead customers... They have made their customers part of the creative process

shows to raise and validate customer awareness. Using a mix of customer labs with groups of around 5 customers at a time, car clinics and innovation labs where lead users or luminaries use and feedback on rapid prototypes, the company quickly evaluates potential design options and configures them around core user requirements.

Participatory Design

Over and above accessing their inputs from a user-needs driven perspective, including customers in the development team to participate in the actual design of a product or service is increasing in use. Several firms now bring together designers, engineers, consumers and internal business clients to participate in co-creation exercises and activities. Variously termed participatory or adaptive design, this is for some an extension of other activities such as using a fan base but for others it is a totally different methodology.

As an evolution from fan bases, several companies have started to make greater use of lead customers willing and enabled to adapt early stage concepts to suit individual needs and preferences. They have made their customers part of the creative process:

LEGO first set up its MINDSTORMS User Panel back in September 2004 to solicit feedback on its existing product which provided a link between the traditional LEGO products and computer controlled robotics. Fuelled by the success of the MINDSTORMS User Panel as an active fan base, in 2006, the MINDSTORMS NXT Development Program drew together users from 79 different countries split into three groups to collaborate with LEGO on the next generation of the product with great impact inside and outside the industry. Having created an active, interconnected customer base, LEGO gained huge benefit from lead user involvement in new product development because, more than anything else, these fans want to be part of the development process. This approach is now core to major co-development activities within LEGO and is being copied by numerous other firms around the world.

Aviva also used a fan base as creative consumers. For the development of the pay-as-you-drive insurance service, over 5000 customers were recruited to use a prototype system. This system was seeking to change the way in which individuals pay for their motor insurance and so gaining first hand feedback from target customers was vital in positioning as well as configuring the product. During an extended trail period, the 5000 participants were able to use the system as part of their day-to-day lives, feedback comments and to also participate in concept development workshops that have changed design parameters to make the system both more effective and more attractive to the users.

Impacts and Implications

As these examples show, successful innovators are making significant and varied use of customer-centred innovation. However, given this wide array of different approaches for improving customer focus within the innovation process, the challenge for many is two-fold: Firstly in choosing the most appropriate technique for the circumstances and, secondly, in actually delivering the greatest impact. The companies highlighted all show that the identified innovation leaders are comfortable with the approaches and their application.

But what about others who may be less well travelled down the innovation journey? For those companies coming to customer centred innovation design for the first time, the primary objective has to be in prioritising what type of insight is most lacking at the moment, where it can be best solicited from and how it will be initially used. With this clear, the choice of approach is usually more straightforward.

OPEN INNOVATION

Five years after the publication of Henry Chesbrough's book, the Open Innovation movement is in full swing. Several firms have appointed Open Innovation Directors, many are venturing forth in multiple directions and conference organisers are queuing up to bring together the leading lights and would be future practitioners. Within this melee, who is actually delivering innovation through an open approach, what techniques are being used and with what impact? This chapter reviews the space, identifies some of the key ingredients for success and suggests how the overall Open Innovation philospy may be taken forward.

Context

Companies have for many years been looking at ways to enhance the performance of their innovation engine. For many, this was primarily driven by greater focus on internal efficiencies of the development process, team structures, decision making and cross functional interaction. However, as internal innovation capabilities became increasingly similar and commoditised, several leading organisations have sought to improve innovation impact through wider collaboration across industry networks and partnerships. Two key drivers here have been in 'filling the innovation pipeline' through accessing more and more new concepts from outside, and 'maximising return on innovation' through finding additional routes to market for internal innovations.

Many of these alternatives were identified in the '80s by researchers such as Eric von Hippel. Whether by using approaches ranging from in-licensing and spin-outs through to co-marketing, firms such as IBM, Xerox, Lilly, GE and Intel have all been doing this successfully for quite some time. IBM allegedly gained $20bn of revenue through cross-licensing via its ITG in 1999 alone and, a year before, Lilly successfully filled a pipeline gap between its blockbuster Prozac and Xyprexa products through in-licensing new molecules from other pharmaceutical firms. Today over 30% of Lilly's portfolio is in-licensed and the success of IBM's cross-licensing has created many imitators. By 2002, as the likes of GSK, Siemens and Motorola all found it more difficult to make their innovation investments pay, a shift was apparent to several industry watchers. As even the toy industry started to get in on the act through the co-development of videogames such as Grand Theft Auto, several academics and consultants were highlighting the increasing trend for internal R&D to be both complemented by external technologies and to offer their own technologies to the outside world.

Henry Chesbrough and Open Innovation

Although Henry Chesbrough reported on rather than invented the Open Innovation philosophy, his name and the approach have become increasingly intertwined. In his 2003 book and numerous other articles, he successfully shared a perspective that several leading firms have been moving from an internally focused innovation process to one that is more 'open'. Bringing together varied examples from Xerox Parc, Intel, IBM and Lucent, he successfully positioned an increasing use of external sources of ideas and technology and external channels to market as a new approach to innovation. His 'Open Innovation' label caught the imagination of a wide range of stakeholders in the growth arena and has since become the mantra for partnered development involving an element of external sourcing, development or market access. The core logic of the Open Innovation concepts has been; that good ideas are widely distributed with no-one having a monopoly; that first to discover is neither sufficient nor necessary for commercial success; that a better business model beats better technology and; that IP is an increasingly perishable asset for which consumer and markets will not wait. Fundamentally, the Open Innovation philosophy articulated by Henry Chesbrough and subsequent disciples, has raised recognition that significant innovation generally happens either in smaller companies or global innovation clusters; that issues such as workforce mobility and venture capital have eroded the ability of corporate R&D labs to contain their useful knowledge and that, simultaneously, a new breed of independent research labs have created a new source of R&D that is in turn being linked through intermediaries to the larger organisations and has thus created an increasingly active and distributed market for ideas. At the same time, recent trends towards increased technology convergence, business model innovation and shortening of product lifecycles have all increased the potential relevance of Open Innovation to many companies.

In-Sourcing

For many, as for Lilly back in 1997, Open Innovation has been largely interpreted as a means to fill the pipeline. The most prominent example of this has been P&G which has now being using 'innomediaries' and internal 'technology scouts' to source external innovation for some time. With the launch of the P&G Connect and Develop programme in 2002, the company sought to better connect its 8000 researchers, 600 partners and 5 business units to realise am ambition of increasing the percentage of innovations sourced externally from a base of 20% in 2000 to 50% by 2008. A core of eighty R&D staff were given the new role of technology scouts to connect into external R&D labs, universities, trade partners and to source new concepts. Supported by parallel programmes such as the integration of on-line networks such as yet2.com and Innocentive, setting up intermediary company-to-company networks like NineSigma, adopting new consumer driven approaches including the use of E.piphany and CoCreate and accessing retirees through YourEncore.com, also co-founded with Lilly, this initiative has made some progress.

The most well known initial product was the SpinBrush, a $5 battery operated toothbrush, originally developed by four entrepreneurs in 1998 and launched in 2000. After successful focus groups, P&G brought the concept and three of the inventors in-house to push fast launch through the Crest brand. Within a year of launch it was the best-selling toothbrush in the US, posted global sales of more than $200m and helped Crest regain the # 1 brand position for oral care in the US. As P&G seeks to deliver an additional $6m of revenue each and every day, it has since continued to push the in-sourcing approach and delivered the Swiffer Duster and Printed Pringles, and established a Connect and Develop hub in Bangalore. Although some P&G technologies continue to be licensed out to other companies and some products are co-developed with other firms, in-sourcing is clearly the core Open Innovation approach that has been embraced by the organisation.

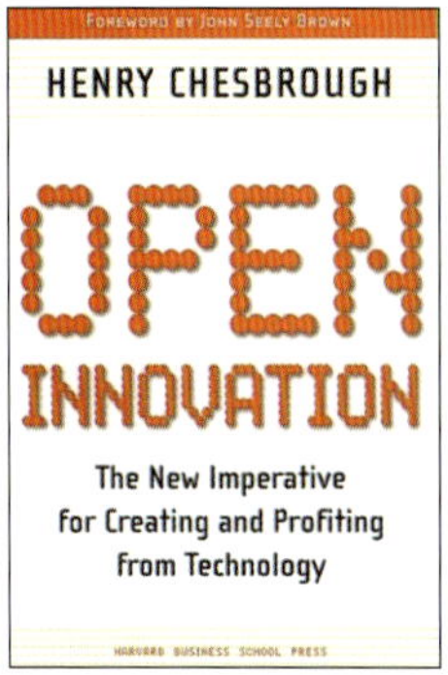

Other companies to have followed a similar approach to using external sources of technology and innovaton range from Novartis and Rolls-Royce to Air Products and Dell. Pharmaceutical firm, Novartis now brings in 50% of its new molecules from a network of university start-up alliances, while aerospace leader Rolls-Royce has strategic technology development relationships with over 25 universities around the world that are now very much part of the company's research ecosystem. Air Products has recently become one of the first industrial gas companies to go the Open Innovation route and Dell now seeks to 'influence industry technology and product directions through sharing customer requirements with strategic partners such as Intel, Microsoft, Oracle and EMC.' Despite its relatively small internal R&D spend Dell is increasingly using its industry leverage to drive others' innovation to market through its brand and delivery mechanisms.

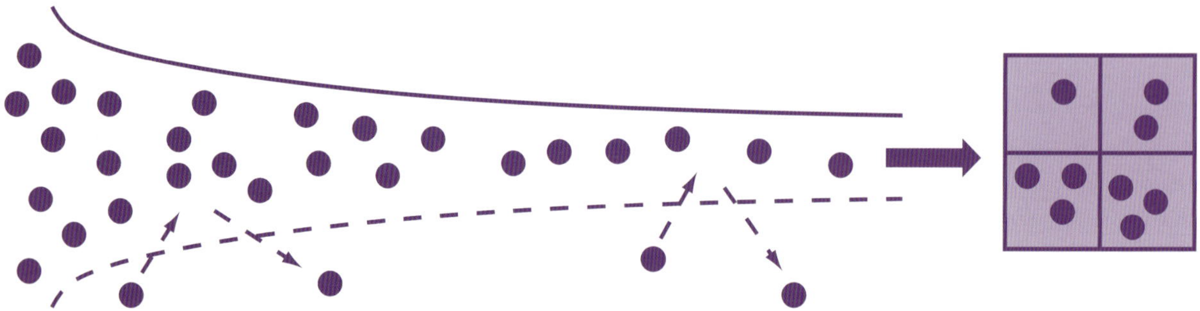

Technology Licensing and Spin-Outs

Licensing technology and associated IP to other companies has been commonplace as a mechanism for generating externally delivered revenues from internal ideas in many sectors for years. For example, with the GSM and WAP protocols, Ericsson and Nokia have been major sources of technology for the whole of the mobile telecommunications sector for decades and, with its CDMA technology, Qualcomm has more recently built a successful business model through licensing to device manufacturers.

Equally, in the IT sector, Intel, Texas Instruments and IBM are all prolific licensors and leaders of cross-licensing to generate additional top and bottom line impact. In addition, there are numerous examples, large and small, of companies spinning out resource and capability as well as technology, often in the form of new venture creation. At one end of the scale, several significant firms built around technologies have been spun out of parents – Lucent was spun off from AT&T in 1996, Thorn EMI spun off its research laboratory as Scipher at around the same time and more recently Motorola spun-out its semiconductor technology and operations to form Freescale Semiconductor. Equally at the other end of the scale, following the lead of universities such as MIT, Cambridge and Imperial College and their associated transfer vehicles which have produced numerous successful start-ups over the years, technology-rich companies such as IBM and GE have also created significant value through moving specific capabilities out of the organisation and into new ventures.

Out-Sourcing

Alongside such technology licensing and spinning out of capabilities or know-how, the out-sourcing of development activities has also been a key part of innovation delivery for some time. Design consultancies such as IDEO have been long-term favourites for external product innovation support for HP, Apple and Samsung - in the same manner as that provided today by software developers such as TCS, Wipro and Infosys in India. Companies work with such organisations either to access capability that they do not have internally, or else to manage gaps in project-based resource. Location and cost have in many cases been secondary issues to capability and experience. At the same time, the growth of Singapore-based Flextronics first as a manufacturer and then developer of electronics has mirrored that of Taiwan-based companies such as Acer and HTC from being OEM suppliers to being 'original design and manufacture' partners.

In a global market, companies are today out-sourcing more and more work to external sources of expertise – but the vast majority of this is still development and not research. Much of it is work that could potentially be done equally well within and outside the organisation, but for which tactically it is more beneficial to move externally. The rest is development that relies on accessing talent and capability that cannot be found inside the organisation – and this is where and why the likes of IDEO and Wipro have made such capital in this space. Together all three of these approaches are successfully delivering on the

Open Innovation allows companies to secure new ideas or find new opportunities to gain benefit from its know-how, brands or relationships

Open Innovation promise for some, but not all. As such they can clearly be seen as key elements in the Open Innovation repertoire.

In and Out

Where the Open Innovation concept has a unique, differentiating impact is when in-sourcing of concepts and out-sourcing of development are linked to create a dynamic vehicle for growth. Four different examples, Apple, Lilly, Philips and DSM, collectively provide some idea of the innovation landscape now in operation:

Apple's iPod is, to some extent, a leading example of Open Innovation that predates the concept - and one driven initially by an external catalyst. In 2001, within eight weeks of an independent contractor first proposing the integrated MP3 player concept, Apple had hired the inventor to develop a product solution and then bring it to market. Drawing together collaborative external resources from the likes of Texas Instruments, Toshiba and Sharp, the technical development was largely put in the hands of Portal Player, a technology development company with an existing platform and the right connections. Linking in with Apple's internal design team and interface developers, the iPod was launched within six months and with over 40m sales to date is already part of contemporary history. The key Open Innovation elements here were the Apple's willingness to both accept an external idea and to respond by pushing much of the development into a collaborative partnership – nearly simultaneous in-sourcing of concept and out-sourcing of development.

Two years ahead of the Open Innovation phrase coming into common vocabulary, Lilly launched the Innocentive marketplace. This on-line forum provides a simple and efficient idea exchange that allows 'seekers' to post problems and the sums – anything from $10,000 to $100,000 – that they are prepared to pay for a solution. 'Solvers' be they scientists, universities and even companies can then offer their solutions. As one of the first commercially focused innovation webs, Innocentive quickly gained a strong reputation as a scientific led marketplace and, by the end of 2005, had over 80,000 inventors working with over 30 major corporations including BASF, Novartis and Nestle. Although arguably pre-empted by the likes of yet2.com, Innocentive is an exemplar of what an Open Innovation ideas exchange is all about.

One of the most significant success stories of Open Innovation prowess is Philips which has changed from being a vertically integrated conglomerate with 18 divisions to a business focused on 3 core areas. Now firmly part of an open ecosystem comprising universities, start-ups and other companies, Philips has used spin-in acquisitions to help it build capability in areas of growth such as healthcare, while simultaneously spinning out technology no longer linked to a core brand, in for example displays. For Philips the bottom line impact of this has been significant.

DSM has been an open innovator for a decade. In the mid '90s it decided to divest its petrochemicals and poly-olefins businesses and re-orientate itself towards the life science, feed, food and pharmaceutical markets, successfully making itself open to innovation across the range. It has made some of its R&D labs the centres of new Open Innovation campuses; through its ventures group it has invested in and successfully commercialised products from small companies operating in such areas as food ingredients; in the past year it has spun out three companies. It has invested in new business accelerators that are targeted with delivering €1bn of additional revenue by 2010. In DSM, although 'Open Innovation' is an external phrase, internally it is seen as an umbrella brand for the real activities - venturing and licensing - that deliver the goods and have been doing so for quite some time.

Lastly, in the public sector, the BBC has successfully developed its own unique approach. To better encourage amateur innovators it has developed backstage.bbc.co.uk to link the wider development community using BBC content and tools. With its own internal R&D activities increasingly focused on delivering Digital Britain and the next wave of digital technologies, this has allowed the BBC to both push incremental development into the lead user base and simultaneously access a global source of new ideas.

Open Source vs. Open Innovation

Several organisations in the IT sector, such as Computer Associate, have been linking the Open Innovation philosophy to the open source movement most closely associated with the likes of Linux. Many see this as the mixing of two quite distinct philosophies with the key point of difference being in the ownership of innovation.

Open Innovation allows an organisation to secure new ideas or find new opportunities to gain benefit from its know-how, brands or relationships. Open Source is, by contrast, driven by the idea that the standards are not owned by anyone and it is through this that greater development freedom can be achieved. Open Innovation uses intellectual property as the core tradable commodity while Open Source seeks to operate outside the constraints of an IP framework. In areas such as the mobile internet, open source and Open Innovation certainly come together but it is a mistake for individuals and organisations to see the two as being interchangeable.

Key Ingredients for Success

External sourcing of technology and generating value through external channels are both significantly different approaches to what most companies are used to, and hence measure performance by. As such, one of the key issues for many has been in defining and integrating appropriate metrics to focus on Open Innovation as a core part of the innovation engine. As more firms have put targets against revenues from new product introductions, several now break this down into core components coming from inside and outside the organisation. Equally some companies have started to put specific budgets against external R&D and numbers around levels of interaction with key external sources of technology such as universities, start-ups and research labs. At the same time, commentators have suggested that companies that have, to date, benefited from Open Innovation have ensured that they provide simple interfaces for people externally to interact with the organisation, reduce the complexity of negotiation by using a handful of simple rules, borrow ideas, create public space platforms for innovation and invest in activities at the boundary of the organisation. More fundamentally, as Open Innovation is essentially about linking external sources and routes to market into internal innovation activities, to find success a company must first be an effective internal innovator. Many are seeing Open Innovation as a substitute for internal R&D, and so the ultimate outsourcing option. This view is the first step to failure as competitors who are better able to deliver innovation through internal activities will win out.

Future Direction

It should be clear that nothing in Open Innovation is really new: The core ingredients have been in place for some time. However, now that most of the innovation world has got Open Innovation on its agenda, the challenge for many is in understanding it. The differences between the P&G type of in-sourcing of concepts, Novartis' access of research through a biotech ecosystem, IBM's technology cross licensing, Philips spin outs and the IDEO / HCL type of out-sourcing of development are still widely misunderstood – especially as they all sit equally under the Open Innovation banner.

The really interesting activities are those illustrated by the examples of Apple, Innocentive, DSM and Philips where external partnerships form part of a dynamic innovation realisation vehicle. If more companies can gain the capabilities, relationships and permission to operate at this higher level, then things could really start to get interesting: The differences between internal and external development will disappear as shared resources become interchangeable; shared IP ownership will become an enabler to faster product introduction and new product and service innovations will drive new perspectives on business innovation. In addition, as more companies follow the lead of Rolls-Royce and big-pharma companies in outsourcing fundamental research through partnerships with smaller firms, universities and independent research organisations, there will be opportunities for even more change.

MULTIPLE INNOVATION

Traditionally many organisations viewed innovation along a singular dimension of developing new products. With the continued growth in services, manufacturing companies at the leading edge of innovation practice are now learning how to successfully integrate different approaches where products, services, processes and business models come together and all drive growth across the board. In addition, established service companies are successfully innovating in the product arena for the first time and establishing a mutual win-win relationship that is increasing margins and revenues. This chapter highlights some of the exemplars that have demonstrated successes as a result of multiple innovation and provide insights into what they have in common as well as what they do differently.

In most developed countries services account for over 70% of GDP and the majority of employment opportunities. As this has continued to rise, an ever-growing selection of manufacturing organisations have looked at how they can develop new services to realize their growth ambitions. While this is nothing new, over recent years there has been a significant shift of ambition. In the old days, many manufacturers sought to offer services to their customers in order to drive increased sales for their core products. As such, few, if any, of the service innovations introduced by product-focused firms delivered any direct revenue; many simply offered advisory services to increase core market share.

Now things are changing - many manufacturers are shifting away from simple product innovation to a more integrated approach that includes innovation in services and processes as well as in new business models. This approach is being led by a few companies such as Apple, Rolls-Royce, Nokia and Medtronic which have successfully introduced highly profitable new services.

At the same time, a number of major service companies have been increasingly embracing product innovation to increase the range of their growth ambitions. Again inspired by a number of key pioneers such as Virgin Atlantic and Starwood, more and more service companies are also looking to increase the cross-over of their innovation focus. Innovation leaders are increasingly adept in the art of 'multiple innovation' – the ability to innovate simultaneously across the whole innovation landscape to deliver integrated growth in revenues and profits. Instead of a one-dimensional focus, more firms are pursuing a two or three dimensional innovation agenda.

However successfully achieving this is far from easy. Shifting from one focus to the other requires careful planning and consideration for differences in approaches to innovation as well as the associated organisational needs and capabilities. In order to improve business performance companies should consider all types of innovation when developing their future growth strategies.

Product Innovators

Probably the best example of a company that has successfully shifted from a product only focus to a multiple perspective, including first services and then new business models, is Apple. With a strong tradition in leading edge product innovation in computers successfully extended into the personal audio space with the iPod, the complementary iTunes service innovated the process of delivering digital music to the iPod and this allowed a new business model to emerge. With a high margin, high volume iPod product range generating significant contribution to the business, the addition of an even higher margin and higher volume service innovation has been a main driver in the stellarsuccess of Apple. Moreover, as the two are interconnected and semi-exclusive, the success of one has continued to drive the success of the other. In addition, as video has entered the mix and the varied cross-over applications in MySpace have gained hold, the whole Apple experience has been very much integrated into the creation of new types of business model innovation.

While Apple's competitors have delivered similar products or services, many have struggled either with ultra thin margins and fierce competition in the consumer electronics sector or the barriers to online music sales. Apple understood the advantages of an integrated approach from the start of the iPod era and this has been central to the way the company has led multiple innovation. The core ambition has never been to be the first but to deliver the best total experience to its customers and make a contribution to the business by innovating in the product, service and business models at the same time.

Nokia has also been busy pursuing multiple innovation strategies. With the 2007 launch of a new service business unit, it has signalled its intent to grow a major service portfolio to complement its sector-leading mobile device and infrastructure product lines. Nokia Music was launched in 2007 and recent acquisitions in the location and mapping spaces have shown that the service innovation portfolio of this traditional product innovator is broadening by the day. As with Apple, underpinning much of the developments is a redefinition of the business model and the relationships between Nokia, its users and the mobile operators through which it has previously established connections to customers.

In other areas, many organisations have identified opportunities for service and product innovation to be mutually supportive in driving growth. However, while many have developed business cases, especially in the healthcare sector, only a few, such as Medtronic, have been successful. As a company known for its medical device implants, such as pacemakers, and a clear focus on product development, Medtronic has shifted to successfully provide complete solutions for chronic illnesses, including diabetes and heart diseases, and is now also moving into real-time patient monitoring, providing critical services to hospital staff.

On a different level some companies are looking to develop different processes to deliver innovation. With the successful 787 programme, Boeing is transforming itself from being a manufacturer of airplanes to becoming more of a system integrator, keeping ownership of the client relationship but pushing out some of the risk associated with innovations. It is achieving this by closely collaborating with partners and suppliers using the latest in IT systems to support the complex design of its aircraft. In the same sector, with its Total Care offer, Rolls Royce has been a leader in delivering not just the airplane engine but in also providing close to real-time in-flight engine monitoring to deliver timely support on the ground where and when needed. The Total Care support services arm of Rolls Royce now contributes over half of the revenues and nearly three quarters of the firm's profits.

Service Innovators

Often new services go hand-in-hand with new business models. When Aviva partnered with IBM and Orange to deliver its hugely successful pay-as-you-drive insurance product, this combination of organisations provided a new integrated product / service / business model into the marketplac. With a GSM device placed in each car being tracked via the Orange mobile network and the associated information processed by IBM, Aviva was able to create a new business model within a mature sector and simultaneously significantly differentiate its products from its competitors. Whilst the core concept had been discussed for some time by a number of its competitors, many of these were focused too much on a one-

dimensional innovation perspective and so failed to see how to best bring the concept to the market. This more integrated approach has delivered collaborative multiple innovation into the market and enabled Aviva to be the first to launch a pilot and quickly grow a new business.

In the travel and leisure sectors, service is usually the main area around which innovation takes place and so both are characterised by ever present incremental developments as one airline, hotel chain or travel agent simply matches or slightly improves upon rivals' offers. However, even within this highly competitive arena, some leading companies are complementing service innovation with innovation in products that successfully make up a total service or customer experience.

Virgin Atlantic consistently, brings leading-edge product innovation into the mix to create a unique customer experience. The company has a deep understanding of consumer motivations and how to satisfy those needs that affect passengers' choice of airline. In Upper Class Virgin Atlantic leads the market and, thanks to a mix of excellent product and service design and its ability to jump ahead of customer expectations, Virgin's reputation continues to grow. After introducing the Upper Class suite across its fleet, the company moved on to focus on its Heathrow Clubhouse lounge which set a new standard for business class travellers.

The latest area to receive an up-lift was the Heathrow check-in experience as Upper Class travellers can go from limo to lounge in only ten minutes – something that competitors can only dream of. In a sector where the core service offer of flying from one location to another is price-matched to and by its competitors, Virgin Atlantic is consistently growing its all-important business class share of the market by delivering a continuously evolving range of product, service, process and business model innovations to its enthusiastic customer base. In a way that rivals cannot even conceive of, Virgin Atlantic understands not only what will make a tangible difference to its target customers, but also how to best deliver this as a key part of a total experience.

Similarly, in the hotel sector, Starwood, has had great success with differentiating itself in the core business traveller market with its Heavenly Bed concept for its Westin Brand. Recognising that many of the mainstream customers for its Westin hotels were looking for a good night's sleep after a hard day, they chose to focus on what was really important. Matching the bathroom, lounge and restaurant experience to that of the core competitive offerings from the likes of Hyatt, Marriott and Hilton, Starwood then focused on delivering the best possible night's sleep by exceeding expectations on the quality of the bed. Unable to find a good enough product on the market, the company led product innovation in this category and created The Heavenly Bed which broke the mould in comfort. Gaining significant uplift in occupancy rates and reacting to the success of the bed the company next starting selling the product on-line. Now, with one of the best selling beds on the market, Starwood extended this

They create concepts that provide a platform on which they can provide combinations of successful products and services instead of thinking of these in isolation

new business model across its other brands with the creation of the W Store where customers can buy much of the furniture, fabrics and bedding so that they can recreate the boutique hotel experience at home. As peers have either embarked on catch-up strategies or stayed stuck in an old paradigm, Starwood is now using the multiple innovation philosophy as a core part of its successful global growth strategy.

Key Lessons

While some sectors, such as those in and around the ICT arena, are arguably more suited to multiple innovation than others, there are many sectors where the opportunities for coordinated, integrated revenue and margin growth that drive product, service and business model innovation go hand in hand.

Even in B2B sectors, many organisations are interested in how to make these services explicit in terms of the value they bring to their clients. Doing this requires a completely different approach to innovation across the board. Organisationally these companies are often geared more towards product selling and sales staff have been accustomed to offering any new services for free in order to win the product business. In order to overcome this legacy, innovating for these companies requires even more consideration about the capabilities needed and how to underpin these with economically viable business models. It many cases it pays to look for stimulation in other sectors which offer different services and operate on completely different business models. Even though they usually cannot be copied directly they can inspire new ways to deliver.

One could assume that there is little difference in how companies approach multiple innovation. Organisations still need to assess their innovation strategy, identify opportunities, develop ideas into products and services and find viable business models that contribute a financial return. However, the approach taken is a major source of competitive differentiation.

Innovation leaders invest in insight-driven innovation that allows them to identify opportunities ahead of and beyond the horizon viewed by their peers.

The more ambitious ones look outside their sectors and current business operations to see whether there are opportunities that could leverage their strengths in other areas. They understand that future innovation revolutions with one space are driven by a mix of changes in adjacent sectors and the rapid acceleration of previously neglected long term scenarios.

Equipped with the necessary insights, they create concepts that provide a platform on which they can then develop combinations of successful products AND services instead of thinking of these in isolation. Furthermore the developed concepts allow these organisations to move beyond their sector more easily and to investigate other business models while considering alternative ways to deliver either by partnerships, collaboration or even acquisitions. The latter brings more than an existing portfolio. It includes both the right culture and expertise associated with the development and delivery of products and services that complement existing offering.

So, whether companies offer products or services, the opportunities of appropriately focused and well delivered multiple innovation are clear. Most organisations have been used to innovating around products OR services rather than products AND services. By considering the full range early on and addressing how they can best leverage one another, more companies could find a path to a more structured approach to multiple innovation that increases their chances of sustained success. The greatest challenge that businesses have to address is to understand how best to include all of these types of innovation in their processes and clarify the most appropriate mechanisms that will allow them to deliver these successfully.

INNOVATION GROWTH VIA SMART M&A

As companies seek to extend innovation activities beyond their core business focus, acquisition based growth is moving into the innovation arena with significant impact. Long seen as the antithesis of organic growth as a large scale activity, smart, small and early M&A creating new growth platforms is now a key element in several firms' innovation strategies.

For many years now there has been a clear view that innovation and M&A are distinctly alternative routes to growth. Moreover, in many circles, innovation as the primary driver of organic growth has been seen to be more sustainable and more successful in building scale and reach than growth driven solely by major M&A. As high profile mergers and acquisitions, including those between AOL and Time Warner, and GlaxoWellcome and SmithKlineBeecham, have variously failed to deliver against their initial targets, other companies in the same sectors such as Microsoft, NTT and Eli Lilly, which have followed primarily organic growth strategies, have largely been more successful in terms of ROI. As M&A failure has been attributed to "poor synergy, bad timing, incompatible cultures, off-strategy decision-making, hubris, and greed", so innovation-driven organic growth has become the primary focus for many firms.

However, as organisations seek to either grow into new spaces of strategic opportunity or extend their innovation capabilities beyond their core businesses, smart, early-stage M&A activity is increasingly being seen as a valuable option for helping to seed innovation growth. Consequently it has become a key component of the innovation toolkit. Some organisations have shifted from the perspective that M&A is primarily about increasing scale and reach to seeing that it is an effective way of building the future portfolio. Either through gaining access to new technologies or key personnel that can become future growth platforms and the champions of this growth, more and more firms have changed their view.

GE for one has undertaken a significant shift over the past decade. Having previously had a reputation for being conquering and aggressive with acquisitions in the Jack Welch era, under its new CEO, Jeff Immelt, GE 'don't buy to grow but grow what we buy'. After years of being seen as a "cost cutting machine", GE is no longer obsessed with the entry price of an acquisition but more focused on the growth potential. As such, it is rapidly being seen as an acquirer of choice for firms in key growth arenas across its diverse healthcare, industrial, infrastructure and media markets. In three months in 2007, complementing its $5bn internal R&D spend, in Europe alone GE closed six deals worth just under $3bn in total – most of which fell under the media radar screen. Many of these new acquisitions are technology focused and based on providing sustainable platforms that can return over 20% annual growth. They are being identified and tracked by an internal team of 200 who are "acting as the eyes and ears of the company".

GE is not alone. One of the primary sectors that has sought to embrace innovation driven M&A more widely has been healthcare. As more firms recognise the growth and margin opportunities available across the medical arena, and so seek to enter the space or build a wider product / service portfolio, smart innovation-focused M&A has become a key vehicle. At one extreme, several companies in this sector have continued to practice traditional M&A for scale and reach. The $27.5bn sale of Guidant to Boston Scientific and the smaller $2.9bn sale of Centrepulse to Zimmer are two recent examples of acquisitions with the goal of increasing market share and product portfolio. However, while these and similar transactions grab the headlines, other much smaller, but potentially more rewarding, deals have been taking place across the sector. Rather than buying established companies with significant market presence and customer base, some medical firms which are focused on integrating major new innovation growth platforms into their business have taken a different approach.

Medtronic is perhaps the current exemplar of what is seen as intelligent early acquisition of small firms to access new technologies, to both grow the portfolio and most significantly, enable broader and higher levels of innovation. With an objective to be 'a growth company for a long-time to come', Medtronic has made investments in R&D and market development which are helping to drive significant growth and expansion of its product range from the cardiac pacemaker products it first launched in 1949. On average, 9% of revenues are being diverted back into R&D, which takes place in 26 research centres around the globe and one fifth of this R&D budget is earmarked specifically for new ventures. While this situation may be found in other companies, a key difference in Medtronic is that, core to its corporate strategy, to maintain a full pipeline and fill gaps in its broadening portfolio the company complements internal organic growth through innovation with focused acquisition of smaller but potentially market-leading assets. In terms of technology-based acquisitions to drive future portfolio growth, 2007 saw both the purchase of Breakaway Imaging to add to its image-guided surgery capability, and the major merger with Kyphon to drive further growth in treatments for debilitating spinal conditions.

Other smaller Medtronic acquisitions in the past few years have included; Image-Guided Neurologics, a company that specializes in precision navigation and delivery technologies for brain surgery; Transneuronix, which is focused on therapies designed to treat obesity by the electrical stimulation of the stomach; Wound Healing Inc, which leads developments in glaucoma treatment devices; Angiolink Inc, with vascular wound closure expertise; Vertelink Corp, providing minimally invasive spinal surgery; Transvascular Inc with a leading position in potential intravascular procedures; Spinal Dynamics, a leader in degenerative disc replacement; and ViaMed Inc which brought capability in minimally invasive prostrate therapy into the Medtronic portfolio.

Although the Kyphon acquisition was larger than usual, a key component in Medtronic's successful growth has been the smart buying of such firms early - after the respective technologies have been proven with lead-customers in place, but before significant market growth has been achieved - and hence before the price to be paid starts to spiral. Led by a team of technology scouts working in close liaison with the business development team, Medtronic has been doing this as a core part of its innovation strategy for some time. As a consequence, and unlike the situation in many of its peers, M&A activity is integrated into technology strategy. There are other organisations such as Philips which seeks to grow its healthcare portfolio through acquisitions, but in contrast to the likes of Medtronic, is largely buying late and, in the eyes of many commentators, expensively – paying way over the odds for more established firms, many of which could have been bought for far less if only they had been on the radar earlier.

Although clearly evident in the healthcare sector, this approach of using intelligent early stage M&A as a key component of building future innovation platforms is not unique to this industry. In several sectors, other companies are taking similar approaches:

In the telecommunications arena, in the wake of Cisco's innovation by acquisition strategy that led to the purchase of 70 companies during the 1990s, many of which did not deliver against expectations, other more traditionally organic growth firms have started on the M&A track. To enable it to broaden the future product and service portfolio, alongside major internal R&D activities, Nokia for one has recently been buying a number of small technology-focused firms. In 2005/6 acquisitions included gate5, Intellisync and Loudeye to respectively enable access to and build future growth platforms in mobile mapping, wireless messaging and mobile digital music services.

In 2007 these were complemented by the acquisition of Twango, a provider of a comprehensive media sharing solution for organizing and sharing photos, videos and other personal media; Enpocket, a global leader in mobile advertising; Avvenu, a company providing secure remote access and private sharing technology that allows users to access and view PC files remotely, and, most notably for a whacking $8.1bn, Navteq, a leading provider of comprehensive digital map information for automotive navigation systems, mobile navigation devices, Internet-based mapping applications, and government and business solutions. These and similarly increasingly high stakes developments, are helping Nokia to integrate ever more functionality into its core products and build a suite of new service businesses that can successfully sit alongside and mutually benefit its traditional product and infrastructure operations.

In the software sector, after traditionally focusing primarily on internally driven innovation, over the past few years Microsoft has also been buying small companies to bring new technologies 'within the tent'. Having bought nine companies in 2004/5 for a total $207m, in the financial year 2005/6 Microsoft bought 23 companies for $649m – an average of less than $30m a piece. These initial purchases of companies and associated intellectual property included MotionBridge, a leading provider of search technology designed specifically for mobile operators and the mobile internet; Vexcel Corp., a worldwide leader in photogrammetry, imagery and remote sensing technologies; Whale Communications Ltd., a leading provider of Virtual Private Network technology for secure remote access to corporate applications and data; Unveil Technologies Inc., a leading developer of enterprise call centre technology; and Giant Company Software, a privately owned provider of anti-spyware, anti-pop-up and antispam tools. In 2007, Microsoft significantly upped the scale of its technology based acquisitions and spent billions of dollars adding companies including:

- Multimap.com - location-based services
- Global Care Solutions - healthcare software
- Jellyfish.com - discount shopping
- Parlano - applications for enterprise group chat
- AdEcn - advertising exchange platform
- ScreenTonic - mobile advertising solutions
- Medstory - health care search; and
- Secured Dimensions - secured software

Early stage M&A is increasingly been seen as a valuable option for helping to seed innovation growth...it has become a key component of the innovation toolkit

In addition, in the same year, the company bought aQuantive, a global digital marketing and advertising solutions company that, at $6bn, was Microsoft's largest acquisition to date. 2008 started off with yet more small acquisitions and then the mega-bid for Yahoo. Clearly complementing internal developments, the need to compete with an ever stronger Google and supporting Microsoft's current innovation strategy, such acquistions are all providing additional capability within the corporation's ever-increasing technology portfolio.

Across the media industry numerous small acquisitions have been taking place. While News International's acquisition of MySpace.com and the $1.7bn sale of YouTube to Google were initial high-profile, high-value examples, smaller acquisitions have been rife: As part of a strategy to grow in select areas in 2006 Google bought dMarc Broadcasting, Inc., a digital solutions provider for the radio broadcast industry for $102m, wiki hosting service JotSpot, Upstartle, a maker of an online word processing program, @LastSoftware makers of SktechUp which allows one to place 3D models into Google Earth, and Neven Vision, a company that specializes in biometric identification. Most of the companies being acquired by Google have small but very talented development teams, innovative technology backed up by strong intellectual property and a relatively cheap price tag. In 2007, alongisde the large $3.1bn purchase of DoubleClick, Google added over 20 more diverse technology acquisitions including Xunlei, a network file sharing company, Adspace, a specialist in video game advertising, Trendalyzer, an analytical software maker, Tonic systems, which focuses on presentation software, Marratech, a Swedish video-conferencing firm and GreenBorder Technologies which focuses on desktop security, as well as Panoramio, a geospatial photo-sharing service, RSS feed specialist Feedburner, GrandCentral, a leader in VOIP aggregation, ImageAmerica which makes high resolution aerial cameras and Zingku, a mobile social network and communication platform. As with Nokia, Microsoft and Medtronic, for Google smart M&A has rapidly become a core part of its innovation and growth strategy.

Key Lessons

For companies seeking to complement internal innovation activities with growth driven by M&A of technically-relevant small firms, there are several points to consider:

- It is the judicial use of a focused team of technology scouts to link into the leading universities and start-ups in priority areas, track progress of new developments, identify the strategic opportunities and buy when the timing is best.
- Supporting this is the ability to see the future early – to have the networks, approaches and strategies in place to scan beyond the current horizon and identify new spaces of opportunity ahead of the competition; and
- Linking all the key elements is the close integration of M&A and R&D strategies around common ambitions that align corporate focus on gaining and building future growth platforms. Having shared clarity on the roles for both organic and acquisition based innovation within the overall corporate growth portfolio is key.

While not appropriate for all, adopting this innovation-driven M&A philosophy to buy smarter and earlier is a valid strategy for some. While many firms focus on delivering higher growth from internally created innovations within their core businesses, firms such as GE, Medtronic, Nokia and Google are all innovating at the edge of their portfolio, filling gaps in their future line up and so seeding higher levels of growth.

LEAP-FROG DISRUPTION

Disruptive, radical and breakthrough innovation continues to be the ambition of many. While some may argue about the semantics, the principle of accessing and delivering higher level innovation is seen as having the potential to achieve higher margins and rates of growth. But what does this mean in practice? What is the ROI on disruption and is it bigger than normal innovation? How can we best encourage and manage disruption? These are commonly asked questions. The challenge is to find informed examples that can answer the questions. Moreover, where there are new disruptions emerging outside the developed world, how can they leap-frog the status quo and where will they have the greatest impact?

Firstly, while a number of technology enabled disruptions based on new science such as stem cells and nanotechnology are having impact, many of today's market changing innovations are actually coming from redefining business models and customer value propositions without significantly altering the technology base. These are relatively 'simple' solutions with fundamental impact.

- Apple's iTunes exploited the unique connection of dedicated supply of software to a leading range of hardware to create a virtuous self-supporting community that is both driving growth in uptake and simultaneously redefining how the music industry functions.
- Skype took the well-developed platform of voice-over-IP and integrated it into a portfolio of services based around the concept of 'free' calls over the internet. By providing free high quality global telephony for users subsidised by connection fees to traditional fixed and mobile phones, Skype took the lead and many of the world's traditional telecommunication companies have been playing catch up as best they can.
- Zopa is an on-line peer-to-peer aggregator site that connects creditworthy people who want to borrow money with a number of people who are happy to lend it to them. Because there's no middleman, both get a great deal. What is more, as the core of the business model, any individual lends their money across at least fifty Zopa borrowers, and similarly a borrower borrows from a group of Zopa lenders. So the risk is equally spread. Lenders get higher interest rates than available through retail banks, borrowers get lower rates than available elsewhere and the banking system loses out.

None of these three organisations is using a specific new technology, nor are they fundamentally changing the consumer experience - listening to music, talking to others, lending or borrowing money. Instead, all are using the design of their business model and the associated redistribution of margins and costs to enable customers to get the service they want for far less than the existing music, telecommunications and financial services industries can provide.

- NetJets is leading the rollout of what is termed 'fractional ownership' providing point-to-point service to fit your time requirements with all the convenience and flexibility of a corporate jet but none of the responsibilities of hiring pilots and scheduling maintenance, and all for a price that is only a slight premium on standard business class – bad news for the likes of BA and United that are seeing some of their most lucrative customers disappear. With over 500 planes, NetJets is an excellent example of consumer value innovation.
- Net-a-Porter provides cash-rich time-poor fashionistas with instant access to the latest shoes, clothing and bags. For a marginal premium on department store prices, customers receive must-have items delivered to the door-step by courier beautifully packaged in stylish black boxes. Run from warehouses around the world with local courier partnerships, the Net-a-Porter business is taking chunks out of the top retailers: Now hailed as the 'first truly global luxury fashion retailer' Net-a-Porter is almost the personal shopper experience anywhere, anytime.

Both of these examples are taking existing business models and proven technology from one sector into another but are doing so, less from a technology push perspective and more from a consumer need pull. The trick for both has been to see the gap and identify exactly what and where is the real consumer unmet need, how it can be best met and at what premium over the alternatives.

Linked into these and of increasing interest to many seeking broader impact is the concept of leap-frog disruption. This is particularly evident where existing technologies are being exploited in new ways in the developing world – where there is the ability to use innovation to bridge the opportunities for disruption and the challenges of addressing the needs and changing dynamics at the 'bottom of the pyramid'. The potential to enable a fundamental shift over a large population and so improve upon the norm in many developing economies is significant. In several instances new innovations are even enabling local changes to leap-frog over the pace of progress in the developed world. Current examples that are gaining significant interest include micro-payments via mobile in Africa and healthcare provision in India.

Whilst many are familiar with the introduction of mobile telephony enabling many African countries to bypass the need to install a traditional fixed line infrastructure to provide near pervasive communication, few fully realise that the same mobile technology may well allow Africa to take the lead in international remittances. Stimulated by Vodafone's successful pilot of the M-PESA mobile money transfer service in Kenya, in 2007 Vodafone and Citigroup announced a mobile based international money transfer service. Users use mobile technology to send money home. Funds can be received by a bank or more commonly direct to another handset and cash can be redeemed at a wide range of outlets such as airtime distribution points operated by in-country mobile network service providers. For the latter, the beneficiaries don't need to have bank accounts, a common feature of many of the world's under-served, but simply have access to a mobile phone that can receive an SMS. The UN estimates that there are 190 million migrants globally with, according to the World Bank, an annual value if over $250bn. Not an insignificant 'opportunity' for mobile phone companies challenged with saturated home markets. Although simple in concept, this service is set to create a true step change in global remittance service – providing money transfer to millions of people for whom a bank account is inconceivable.

Aravind is an Indian healthcare company that, amongst many other services, is pioneering low-cost cataract operations. Aravind Eye Hospitals have been running for 25 years and over that time have sought to address the challenge of providing traditionally expensive surgery in developing countries. Through redesigning both the cataract operation and the supporting procedures, Aravind can perform a $500 operation for around $50. In addition, as a result of a unique fee system Aravind is able to provide free eye care to two-thirds of its patients from the revenue generated from the other third of its paying patients. Not only is this company providing millions of patients with healthcare that was previously inaccessible but the processes they have developed are also having impact in other economies. As the concept of medical tourism evolves from 'nip and tuck' vacations in Brazil and South Africa into dental treatment weekends in Hungary and cardiac surgery in Dubai, so travelling to India for an efficient and timely cataract operation is now also

becoming a viable economic option for many in Europe and the US. What started out as a low cost approach to a much needed operation in India is now shifting the economics of supply in the developed world.

These two truly market changing innovations are integrating the adoption of new technologies through new business models to improve the local customer experience and so, in many ways, are having far greater relative impact than those such as Zopa, Skype and NetJets mentioned above. Not only are they touching many people, they are also enabling a significant shift locally and also providing new models that in many ways are more effective or efficient than that which is available in the US and Europe. Whilst the technology could have been implemented in the developed world for years, and in the case of mobile micro-payments has been talked about by varied telecom operators and financial institutions for a decade now, the very fact that existing infrastructure, service providers and commercial relationships existed has, in the eyes of some, been the largest barrier. The need to align the interests of so many stakeholders, some of which would gain revenues but many of which would lose current business, has been a challenge that few have been able to overcome. They have been unable to overcome the inertia of the legacy systems. In fact, even where prototype systems have been rolled out in such countries as Finland, Japan and Hong Kong, they have largely only occurred because of exceptional, and often government facilitated cooperation.

Equally, the process and hence cost of providing cataract operations could well have been redefined in the US years ago, but the interests of the existing healthcare providers, insurance companies and government bodies had created an environment tuned to incremental innovation and so one that has not been predisposed to the potential of leap-frog disruption.

What should be of concern to some, and reassuring to others that are already involved in enabling major change, is that there are many more potential leap-frog disruptions currently in process.

Chinese conglomerate Huaneng is cooperating with Tsinghua University to make the concept of pebble-bed nuclear reactors reality. These small, efficient and, significantly, intrinsically safe energy sources are being planned to meet the rising energy demands of many of China's cities that are not linked to major fossil-fuel supplies and in doing so will provide a clean secure source of the country's future energy demand. Each reactor uses around 600,000 spherical graphite fuel and moderator elements as the core, nuclear waste is controlled and power of up to 300MW is easily delivered. With China as the 2010 launch base and the rising worldwide awareness of the benefits of nuclear energy as a clean and secure future power source, the potential for Huaneng globally is evident. It won't be long before this organisation is providing solutions back to the developed world in the same way as Lenovo will in the PC market.

The trick for both has been to see the gap and identify what and where is the real consumer unmet need, how it can best be met and at what premium over the alternatives

Also in the energy field, the opportunities to be gained from widespread adoption of distributed power to provide electricity to rural areas of India, China and several African states where current power infrastructures are either absent or inadequate, could well drive acceleration of the development and roll out of low-cost efficient solar cells and implementation of new micro bio-mass systems at a village scale. The local relative cost of supply and the lack of credible sustainable alternatives means that absolute cost is a secondary issue and, as scale is guaranteed, overall economic benefit can be quickly realised.

Thirdly, the increasing demand for potable water in many regions is renewing focus on the opportunities from desalination systems that can effectively clean dirty water. Through the development of new membrane technologies that could for example be based on bacterial systems, whichever country that takes the initiative and makes a breakthrough could well define a cost-effective, pragmatic solution to the inevitable water scarcity that will affect significant areas of the planet over future decades. Several countries are already investing heavily in the key research required to adopt and adapt largely existing technologies to the growing demand for increased access to potable water for a growing population.

Given the evident local needs, the increasing access to enabling technologies and the clear market potential, innovations such as these will provide major future disruptions driven by local needs and development of scaleable solutions.

In a number of other instances, developed world know-how, rather than capital alone, is itself enabling the developing world to build new solutions that exceed most developed world norms. One excellent current example of this is the design of Dongtan, a new city just outside Shanghai, by the leading engineering firm Arup. As 600 million Chinese move from rural to urban environments by 2050, the demand to build new cities that exhibit far higher levels of sustainability is increasingly recognised. Using a unique planning methodology termed integrated urbanism, Arup have designed an 8400 hectare city that is a mixed use environment that redefines conventional wisdom of what a city should be. Dongtan will have less than half of the typical ecological footprint of a traditional city: For example, water consumption will be down by 43% and water discharge by 88%. In addition there will be a 64% reduction in energy demand with no carbon emissions. The thinking behind the design of Dongtan is not only breaking the mould of how cities are conceived but is also serving as a model for many other Chinese cities to similarly leap-frog the standards of the West. Moreover, many of the design principles being executed on this project are now being re-imported back into Europe and the US in such urban developments as London's Thames Gateway project. Driven by Arup's design prowess, but led by the foresight of the Shanghai authorities, this is one of several examples of Chinese 'catch-up' fast becoming a leap-frog opportunity.

Successful disruptive innovation relies on matching unique insights around both new and existing technologies with clearly significant customer needs at a global as well as local level. The challenge for many wishing to enable, deliver and profit from leap-frog disruption is in having the insight and foresight to see where and when the most significant opportunities for sustainable change will occur. What will they be, what will be the core drivers and what will be the likely impact? How some are doing this is covered in the next chapter.

IDENTIFYING FUTURE GROWTH OPPORTUNITIES

With the pressure to deliver yet more growth impacting most organisations, many companies, especially those seeking major breakthroughs and disruptions, are looking way beyond their traditional horizons. In order to identify major new spaces of innovation opportunity that may offer substantial prospects for sustained growth, many are also adopting new approaches that mix foresight, insight and cross-sector interaction. Those that are successfully using these approaches are finding the rewards to be highly beneficial as they not only see future opportunities and threats ahead of their peers but can also put in place the necessary projects to seed the potential growth prospects.

As the world around them changes, many companies today still fail to look over the immediate business horizon to identify the key challenges and major growth opportunities that are potentially available. Frequently having no clear vision of the broader future consumer or technology scenarios that lie ahead, these organisations often display poor integration of growth strategy within the business - particularly between R&D and commercial functions. In addition, they are typically over-reliant on seeing the future as linear extrapolations of today and so miss tangential opportunities. As a consequence many focus primarily on incremental developments with fewer breakthrough innovations.

Alongside this there is frequently a closed attitude towards externally sourced or cross-sector innovation and little recognition that today's consumers can successfully help inform decisions about the future. As a consequence of all this, these companies consistently underperform when compared to innovation leaders, and have a corresponding low return on innovation.

Faced with increasing commercial, stakeholder and consumer pressures to innovate at higher levels and so identify the key growth opportunities that lie ahead, many global organisations need to better understand potential futures. The best corporate strategies are built to accommodate different scenarios and, within this context, although some economic and social trends are relatively easy to predict, reliable technology and consumer scenarios are more difficult to establish.

Technology and consumer futures are inherently disruptive in nature, prone to unexpected events and, unlike economic and social forecasts which can be reliably accessed from the usual sources ranging from the UN to Goldman Sachs, they are less likely to be simple extrapolations of the here and now. While each sector has its own view of the future, commonly known as foresight, the reality is that it is the convergence of different technologies, markets and communities that creates real change. Unfortunately for many companies, identifying this ahead of their peer groups is no easy task.

Although SBC, Cingular and Motorola have a clear view of what could potentially occur within the telecommunications arena, Pfizer, GSK and Boston Scientific understand future healthcare developments and Ford, Siemens, Nissan and Honda variously have views on the future of transportation, few companies truly have an in-depth understanding of the opportunities that will develop between the sectors. For instance, it is only really the likes of Nokia, Medtronic and Toyota that have views on how the telecommunications, healthcare and even automotive worlds may actually come together to provide continuous health monitoring and diagnosis of a population that is often in transit.

Likewise, there are relatively few organisations that have a good steer on how the sources of key sensing, location and micro-energy technologies may cooperate and integrate with emerging consumer needs to realise the potential from product, vehicle and even personal tracking. Those that do can deliver fundamental changes to such areas as insurance, leasing and logistics and so modify the relationships between manufacturers, customers and other stakeholders.

Investment in new technology especially has to be guided by clear insight and by seeing how technologies will interact, conflict, create and enable new innovation and thus impact our world, is a key gap for many companies. Accessing this sort of insight is not easy. As mentioned previously, there is a tendency in many areas to overly rely upon the linear extrapolation of existing trends that can miss tangential opportunities. There is also often a limited breadth of technology and consumer expertise within individual organisations and few forums exist for cross-sector interaction with external experts. In short, in order to place well-informed intelligent bets on the future, corporate and government organisations alike increasingly want to differentiate between possibility and probability but there are few truly reliable ways in which they can do so.

More and more companies are recognising that while in-sector perspectives are valid, they are not as rich a source of future innovation insight as well-choreographed cross-sector views that bring together a rich mix of leading sources of insight from multiple arenas. If these are developed using truly global leaders in their fields who have already evaluated and filtered a wealth of the latest insights and are actively synthesising deeper knowledge, then harnessing this highly experienced knowledge pool, can jump-start programmes straight to the leading edge of the debate across converging sectors. Moreover, while single sources of expert advice can be very useful when the scope of focus is narrow around, say, a single technology platform, consumer segment or market, to examine multiple divergent and potentially disruptive options, debate and discussion between leading experts is a far more effective means of exploring the unknown and identifying unique opportunities.

Today there is a select number of leading organisations such as Shell, IBM, Siemens and Nokia who have either undertaken or are in the process of starting to gain clear pictures of the future through differentiating between possible and probable impacts, events and opportunities. With support from a number of key institutions, they are defining future innovation strategies by variously using participative customised foresight techniques to engage internal and external experts in shaping robust views of the future; highlighting break-through opportunities that lie at the intersections of emerging technologies, markets and consumer needs; and connecting and developing consumer, market, societal and technology trends through joint opportunity road-mapping. As a key part of this, some are involving the consumer as a

fundamental source of innovation foresight and successfully linking insight and foresight across functions, activities and programmes. Although outputs and the ways in which they are used by the organisations differ from one to another, the programmes that they are undertaking are producing engaging validated strategic scenarios to establish a preferred company future viewpoint and they are all delivering outcomes that can be communicated company-wide to inform innovation and growth programmes.

Technology Futures

Technology Futures programmes provide unique views. Built around specific and challenging questions that are designed to address areas of future uncertainty, they bring together the best minds from a number of fields, and debate and map out a rich insightful perspective of the future. They create detailed views of both impacts and applications, and

highlight the probable sequence of future technology development. They also identify pivotal technology and policy choices that can lead to alternate paths and options and clarify the key enablers, inhibitors and uncertainties. Two of the currently most recognised Technology Futures programmes are undertaken by Shell and IBM.

Shell's Technology Futures programme is run every three years on a schedule to fit in with the organisation's well known Global Scenarios. Bringing together leading sources of technological and social insights from across multiple sectors for week long events, it looks twenty years out at a broad range of global themes – everything from 'healthier living' and 'always connected' through to the 'future of business' and 'small and distributed'. The Shell Technology Futures programme is seen by many firms as the benchmark in future growth opportunity identification. The first programme was run in 2004 and successfully identified a number of major future business opportunities for Shell. The programme was therefore repeated in 2007 and is now a regular part of the company's strategic input.

IBM's annual Global Innovation Outlook uses the same structure as the Shell Technology Futures programme but with different focus and timescale. Bringing together academics, clients and partners of IBM it looks at a five to ten year timescale and focuses on just three themes at a time. The first programme, run in 2004, had over 100 external contributors from 96 organisations mixed together with 100 IBM researchers and focused on the topics of 'healthcare', 'government and citizens' and 'the business of work and life'. The second GIO in 2005 focused on 'the future of enterprise', 'transportation' and 'the environment.' Now established as a annual strategic insight activity to complement internal perspectives, the GIO is an important part of the IBM innovation portfolio.

Such programmes produce detailed, validated outputs that support an organisation's long-term strategy, inform major investment decisions and also guide future R&D or competency development. These are fundamentally different from traditional forums. Whereas other multi-sector events have hundreds of participants, most of whom have brief interactions around presentations and seminars, Technology Futures programmes involve sustained interaction, debate and discussion between smaller groups.

While other forums are based around broad themes, often involving corporate leaders and figureheads and are largely focused on personal and corporate networking, Technology Futures programmes use interdisciplinary interaction between the real sources of future insight to address specific questions. Other forums are often openly reported, focus on a consensus of topics and provide transcripts and declarations as output. Technology Futures programmes are based on confidential discussions about the impacts and opportunities of new technology where the host organisation influences the overall agenda and focus, and the outcomes are

To place well informed intelligent bets on the future, corporate and government organisations increasingly want to differentiate between possibility and probability

actionable technology impact maps and narratives. They thus create an effective, accelerated platform for the subsequent and confidential exploration of specific scenarios and opportunities. In particular, they regularly identify major new spaces of major innovation opportunity or market challenge that can potentially lead to the identification and development of significant new growth platforms for the businesses concerned.

Consumer Futures

In other sectors, companies require greater understanding of key consumer insights and trends in their medium term innovation programmes and so have recently been adapting the Technology Futures approach to the consumer agenda. Consumer Futures events are now helping companies to define future consumer behaviours, drivers of change, emerging new products and service needs as well as potential market disruptions. Again focused on the bringing together of a unique mix of leading sources of insight from non-competing organisations within an intensive, immersive discussion, Consumer Futures programmes identify unique cross-sector intersections of innovation opportunity and enable companies to clearly map consumer foresight landscapes beyond the traditional 3 to 5 year horizon.

Recognising the need to be more attuned to further-out potential consumer changes and understand the value of deeper, more informed insights sourced from outside their home sectors, firms across the financial services, FMCG, food and media sectors are all now running Consumer Futures programmes to identify major new business opportunities. Such organisations are often mapping the consumer landscape in the 5 to 10 year future to create clear and compelling pictures of what this may mean for their brands, products and touch points with consumers.

Finally in a couple of companies, Technology Futures and Consumer Futures programmes are being further integrated to identify other unique opportunities that leverage both fields. Following similar ambitions realised by organisations such as the BBC and Nokia, both of which have ten year views of how the broader range of digital technologies will drive and enable new consumer behaviours and what corresponding product and service innovations will be required, companies in the FMCG and electronics sectors are exploring new intersections between technology, society and consumers. Using a common approach, they are identifying breakthrough platforms upon which new consumer benefits can be delivered through the application of emerging technologies across their brands.

Driven by increased shareholder expectations for sustained long-term growth, customer expectations for continuous innovation and internal ambitions for greater impact, it is clear that the adoption of such futures programmes will soon become part of the innovation portfolio of many leading firms. However as they all seek to identify unique growth opportunities, the challenge for many will be in out-manoeuvring their peers who are also adopting similar approaches. Speed will, as ever, be a key differentiator: While two competing companies may well access the same insights, it will be the rate at which they can act upon the insights to guide R&D and brand strategies that will be a critical issue.

At the same time, as more and more companies such as the Institute for the Future, Philips, Shell, Siemens and IBM all openly share many of the insights around the future that they are gaining, a key capability for others wishing to leverage them will be in the simultaneous adoption and adaptation of the key lessons. More generic probable views of future possibilities will become more accessible, but identifying the distinctive growth opportunities for an organisation may well become an increasing challenge.

FUTURE INNOVATION CATALYSTS

Building on the last topic, this chapter looks at some of the key catalysts for innovation across markets and sectors. Drawing on the insights gained from multiple futures programmes, it highlights how catalysts are drivers of innovation, identifies those that are currently having influence and then provides an overview of three of the key future innovation catalysts that we see as being of note for all.

As innovation continues to migrate across sectors, markets and technologies, companies are increasingly interested, not only in specific innovation opportunities, but also in the core drivers for future innovation. Moreover, as highlighted in the last chapter, more organisations look to explore the potential to be gained from process and business model innovation. Alongside the more traditional product and service innovation, they want to identify where and how these core drivers will have impact and how they can best be exploited. For some, these drivers are seen as potential catalysts for major innovations and are therefore topics that need to be understood so that the implications for their markets and evolving capabilities can be explored in depth.

The core catalysts for innovation today are easy to identify:

- Carbon and the broader green agenda can be seen to be driving innovation everywhere from cars and food to financial services and housing;
- Web 2.0 and the associated changes to internet content are at the forefront of many a media company's innovation agenda;
- The end of oil and the rise of nuclear and bio-fuels is having impact not just in the energy field but also in land use for food production, chemicals development and government policy; and
- Wellness and the challenges of keeping an ageing, more obese population active and healthy for longer is a key driver of new ideas across the food, drink, leisure, FMCG, pharmaceutical and broader healthcare sectors.

Although some organisations fail to fully understand the detail and likely implications of these down the line, most innovation directors have a good working knowledge of the general issues around them - and hence the potential innovation opportunities that they present.

However, as companies look forward to increased revenues from innovation and so wish to identify major future growth opportunities, few are currently really grasping the challenges of identifying, scoping and accommodating the likely future innovation catalysts that will make and shape the next generation of new products, services, processes and business models. Some look at emerging and accelerating generic trends that are often extrapolations of today as sources of potential insight; as yet, only a select few are specifically addressing the need to understand how and where major changes will impact their industries and markets - and what will be the associated key future innovation catalysts.

As firms like Innovaro continue to run an increasing range of futures programmes for companies across many sectors, they help companies see future opportunities both within and around their current and envisaged centres of activity.

In 2007 Consumer and Technology Futures programmes for companies in the energy, food, automotive, FMCG, financial services and materials sectors touched upon and linked into ongoing developments across everything from biotech to new communities. From this some of the key future innovation catalysts are clearly visible. Three of the catalysts that are key for the future, either in terms of probable mass impact on society or more as cross sector drivers of innovation are water, authenticity and healthcare insurance.

Water

Foremost in many future agendas is the growing problem associated with global and regional water scarcity. As the world's population continues to increase and migrate into urban areas; as the demands of economic growth in India and China drives broader industrial growth and as average freshwater consumption per capita in Europe and the US also rises, it is clear that the current trajectory is not sustainable and radical changes are required. Within a decade wars may be fought over water rather than oil as water is seen as the world's most precious resource and so, recognising the scale of the challenge, a number of insightful organisations are already examining what can be done to offset and exploit this. On the supply side, key regions of the world such as Singapore and the UAE are investing heavily in economical water desalination technologies that can convert salt water to fresh water, while companies such as GE are examining new business models to create and deliver alternative water resources for agricultural and industrial use. Equally on the demand side, future washing machines and dishwashers are being designed to run on a fraction of today's water consumption; FMCG companies are looking at shipping smaller, more concentrated product; and architects are focused on grey water reutilisation alongside rain harvesting techniques. Water is the next big challenge for society and will certainly drive innovation across the board, especially when the carbon debate also moves forward and, for example, manufacturers start to label their products with the amount of water used in their production, not just as a regulatory requirement, but also as a means of purchase decision differentiation for the customer.

Authenticity

Knowing what is real and what is not, and so what has value and what does not, is an issue that will impact a wide range of sectors. Although the topics of counterfeiting and anti-counterfeiting detection have been main-stream in consumer areas such as bank notes, razor blades, CD / DVDs, handbags and sunglasses for years, they are already a major concern in such diverse arenas as pharmaceuticals, aircraft spare parts and paint manufacture. Even as China and India start to align themselves with international intellectual property protocols, knowing which products are authentic in these countries and the wider marketplace is already a major issue. As we move forward authenticity concerns will spread wider. For example, as user generated content becomes the norm rather than the exception in mass media and the 'pro-am content' environment takes hold, who will we trust more for authentic information - audience input to BBC and CNN or wikipedia? Vastly increasing speed and range of supply of information, opinion and supposition will make authenticity a core issue for us.

Equally, authentic identities will become a major issue. As DNA based identification supersedes biometric passports and cards being introduced in the near term, will access to information, buildings and countries require higher and higher levels of authentic security – or less? Also, in terms of capital investments, what will have authentic value in the future? Many commentators have been surprised at the values companies have been attaching to islands in second life as major brands seek their own space in the virtual world. But there is a strong parallel happening in the real world – countries, such as Dubai, which are building multiple artificial islands, are extending what was once a limited resource market of beachfront property into a manufactured product where another 1000 properties can

be created from scratch. In such a world what is an authentic piece of real estate, and what is not, will either become a source of distinction or else the values associated with them will have to stabilise. Some will not care what is real and what is not, but for many the authentic information, product or even space will be a much cherished item.

Healthcare Insurance

Lastly, we come to the increasing role of the healthcare insurance industry in setting the framework and defining the macroeconomics for a wide range of future business innovation. Just as real-time vehicle tracking is enabling pay-as-you-drive motor insurance to pave the way for a broad range of new innovations in the transport and telecommunications sectors, from intelligent highways to car-to-car networks, so over the next ten years healthcare insurance will influence a growing range of issues. Early signals of this include Swiss Re setting a limit on the body mass index of people they will provide reinsurance for as a means of excluding heavily obese people from being customers for core health cover and Prudential's discounted health insurance linked to attendance at a gym. The ability to use insurance cover and premiums as both a carrot and a stick to drive customer behaviour is not new but going forward healthcare insurance companies will become a major driver of other services.

Close to the core, real-time health monitoring is a much discussed area of innovation opportunity, but many companies still see this as something that customers will either pay for directly or that hospitals will provide. For a few lead users this may be the case, but for the mass market it is not credible. Proactive consumer purchase and use of vital signs monitoring is a niche and will remain so, and few, if any, hospitals can afford post-therapeutic support of this order. It will be the healthcare insurance companies who will be able to drive mass adoption as they offer variable premiums to customers who firstly link into the system and secondly change their diet and exercise regimes to improve their health. Most likely delivered directly via mobile communications for continuous connection, real-time healthcare monitoring will become commonplace in the not too distant future. More on the edge of current practice, others see healthcare insurance companies also being key drivers of innovation in such areas as future food design – both in terms of composition and portion size, employee and consumer focused eco-footprint schemes and in-car driver alertness indication.

These and other future innovation catalysts should be understood by all Innovation Directors. Whether core or peripheral to current areas of focus, down the line they will have impact on many of today's businesses. Having the insights to assess which of the catalysts will impact where and within what timescale is a first step for many. Clarifying what the implications are is the more subtle second step. Translating these into future growth strategies is an area where, to date, only a few leading companies have really made the critical final step. However, as future innovation growth becomes increasingly linked to and driven by the understanding of key catalysts, more organisations will have to follow.

Future
NEXT EXIT

MEASURING INNOVATION

While organisations increasingly seek to profit from innovation as it is seen to be the primary driver of sustainable organic growth, so they and varied interested parties want to more accurately measure it. In a book focused on profiling the companies who are the leading innovators in their respective sectors, many readers will want to understand more about how to measure innovation. However given the range of levels within the organisation that innovation performance can be tracked, the varied philosophical approaches that can be adopted, the number of sectors that are trying to find a common language and the different types of innovation that need to be included in the mix, gaining a clear view of true innovation performance by selecting the most appropriate metrics is complex.

Whilst in some circles innovation has always been viewed as one of those intangibles that cannot readily be accounted for and so, in terms of ultimate business contribution, cannot be reliably measured, in others attention on getting a better handle on what tangible elements exist has been growing. In fact, after years of being a minority interest for many, these days it seems as though everyone wants to measure innovation:

- Companies want to better benchmark their innovation performance against peers and they increasingly need to monitor internal progress;
- Academics want to identify key metrics that inform the evolution of major innovation themes ranging from collaboration and Open Innovation through to IP exploitation;
- Government departments from education, science and industry to healthcare and transport all want to track the innovation and economic impact of the funding and investments they have been making to companies and regions;
- Countries as a whole want to measure their innovation capability and associated economic competitiveness as they seek to drive growth and attract inward investment;
- Analysts want to understand which companies are delivering the most effective organic growth from their investments and which will grow best in the future;
- Consultants want to integrate innovation measures into overall company performance audits or push their own views on which metrics matter to win business; and
- Software providers want to persuade companies to track certain metrics as part of products to automate innovation processes and map innovation portfolios.

It's all getting mighty crowded and, for many, very confusing! As such, a number of organisations are seeking to either highlight the common issues and associated measures that work for all of these circumstances or alternatively identify the unique metrics that apply best to their own individual position, ambition and strategy. To help make some sense of the multiple options available, below we have outlined some fundamental issues that many companies are, to varied extents, successfully tackling as they improve the measuring of innovation within their organisations.

Corporate Metrics

Firstly, at the highest level, the vast majority of companies now align their innovation activities around delivering organic growth and so use a revenue growth target as the primary means of setting the innovation ambition, measuring associated progress and reporting subsequent success. One high-profile example, P&G, is seeking to add $4bn of additional revenue a year which equates to mid-way between its declared 4% to 6% organic growth target. Likewise, DSM is looking to add €1bn of revenue from innovation, while Nestle's organic revenue growth target is 5 to 6% and Nokia's is 20%. For some of these companies, such revenue growth can be partly achieved by increasing sales of existing products in new or existing markets but mostly it can only fully be delivered by the introduction of new innovative products and services. In other organisations such as Philips and BT, which, for different reasons, are not succeeding in significantly growing revenues, the focus is more on margin growth and so this is driving a migration of product portfolio from traditional to new areas of innovation action. However in both cases, this ultimate ambition of achieving growth is common as the core outcome of the majority of innovation activities.

Business Metrics

Underneath such high level targets, the options for innovation measures within individual businesses are multiple and varied and, to complicate things further, different organisations have markedly different views on if, how and where to adopt innovation metrics.

At one extreme, several companies ranging from Amazon down to many small start-ups don't measure anything specific below top and bottom line growth. For these organisations innovation is the key means of achieving growth and so is the vehicle by which corporate growth targets can be met, stock prices can rise and shareholders can be rewarded. Top and bottom line growth ambitions are the primary focus for these organisations, they drive behaviour and their results are how individuals, teams and business divisions are rewarded.

At the other extreme there are several companies which have a detailed cascade of innovation and growth targets embedded in the organisation. GE is a well-known and well-documented example at multiple levels. BP uses individual performance targets to help deliver its corporate aims across a wide range of areas, including innovation, while household products manufacturer Reckitt Benckiser is increasingly recognised as being exemplary in using innovation metrics across the organisation to help drive behaviour and so maintain continued rising business performance.

Both of these extremes work for the organisations concerned. They fit the respective cultures, drive desired behaviour and support their innovation ambitions. However, for the vast majority of companies, which operate between these extremes, there are a wide range of options being used.

Why and What to Measure

What innovation metrics are being used by an organisation is, in many cases, frequently driven by why measuring innovation is seen as being important. For example:

- In some companies, especially those seeking to introduce new innovation practices, the onus is more on tracking but not necessarily rewarding innovation performance. As different approaches are experimented with it is more important to understand which work, and why, rather than to specifically reward their use and the associated impacts.
- In some firms, especially those with a high level of technology input, there is often a focus on measuring and rewarding innovation projects as a whole, rather than individual contributions. As such the key interest is in meeting market, technology or consumer targets.
- In many service businesses, where the process by which new products are delivered is the focus for much innovation effort as opposed to the products themselves, innovation is frequently linked to efficiency and so this becomes the primary focus for metrics.

All three are clearly different situations and so require different measures to be put in place, used in different ways and variously linked to higher level corporate targets. Moreover, as companies grow the reasons for innovation often evolves, so too the typology of metrics that are adopted.

Next in the metrics maelstrom comes the option of what to measure. Across the innovation landscape there is a wide range of different business / project / product and individual metrics being used. This is at one

level a fundamental problem because it is the very lack of a suite of generic measures that works for all companies that makes it so challenging for one company to be compared to another. On the other hand, it is in itself a reflection of the complexity of the innovation arena and so, for those that know what to look at, more revealing in the detail. In reality there is a smorgasbord of options being used.

For most governments and many ex-public institutions such as utilities, telecommunications providers and transportation firms, the primary focus is often on measuring the input to innovation. The obsession in some areas on R&D investment is a primary example here. On the assumption that the more that is spent on innovation, the more innovative a country / region or company will be, this has become the singular focus for many. Indeed in some areas, as R&D intensity becomes the primary issue little attention is paid to what impact the money delivers.

For some companies however there is more interest in the output from innovation activities. Whether this is the number of ideas, number of patents or number of product launches, this is an easy way of looking at, and comparing year-on-year, peer to peer performance. When embracing such measures as EVA it can include the concept of added value and if, for example, tracking how many products include external input, can even accommodate Open Innovation requirements. However, like input measures, when used in isolation output metrics are increasingly seen as an ineffective way of gaining a realistic view of the true effectiveness and efficiency of innovation within the organisation. Input and output needs to be somehow related.

Several organisations are more concerned with the effectiveness of innovation - the impact or outcome of their innovation activities. On the basis that it is the impact of innovation rather than the number of innovation activities or projects underway that most significantly effects business performance and the contribution to growth, firms such as 3M have for years looked at the revenue generated from new products launched in the previous 1 or 3 years. Other companies look at change in product profitability, market share, brand value or customer satisfaction.

For more sophisticated firms focused on improving innovation performance, assessing the efficiency of innovation can take multiple forms. At its simplest level, companies look at project indicators such as time to market, time to profit and resource utilisation or business indicators such as, in the food sector, the percentage of the product portfolio that is on the market for more than three years. At a more evolved level, organisations such as HP have for years been tracking the return on innovation while others are now increasingly seeing merit in some of the efficiency ratios that underpin the annual Innovation Leaders analysis such as average R&D spend per patent, increase in revenues per employee and resulting change in margin per product or customer.

The more enlightened organisations are currently taking two alternative routes; for

The challenge is not in finding candidate metrics to use, but rather in choosing the most appropriate

some, where innovation is already in the DNA and is successfully delivering sustained year on year growth, the use of metrics is actually declining. The innovation engine is largely running well and, so to ensure that it keeps going, only a few key indicators are used for fine tuning. For others, mainly those that are seeking to effect more of a step-change in innovation performance, a number of key metrics are finding favour either as temporary means of getting things on track or as part of a longer term innovation balanced scorecard where external, internal, financial and future growth indicators are all integrated into a cascaded scorecard.

Sector Differences

The above notwithstanding, there is then another complication that some business leaders fail to consider. This is the fact that innovation in one sector, and hence the indicative measures that can be used, can vary widely from innovation in another sector. Whereas patents are indeed a highly appropriate measure of innovative activity and future revenue potential in the pharmaceutical sector, they have less relevance elsewhere. Equally while many service firms such as mobile phone operators and retailers are interested in increasing average revenue or margin per customer as the primary outcome of new innovation, for other service firms, and especially those with a strong internet channel where cost of delivery is nominal, such considerations are secondary to audience and reach.

Different Levels of Innovation

Lastly there is also the issue of the level of innovation that is expected, being delivered and hence needing to be measured. Whilst many aspire to achieve breakthrough radical innovation, where appropriate measures may range from brand impact and customer awareness for product innovations through to decrease in cost of supply or transaction for process innovations, the vast majority actually make money from delivering incremental innovation where such measures are potentially less applicable. The success of incremental innovations such as line extensions and portfolio repositioning is, for instance, often measured in terms of protection of margin and market share.

Conclusion

Given that innovation is widely seen as an increasingly important business capability, the challenge is not in finding candidate metrics to use, but rather in choosing the most appropriate. There are tens of innovation metrics now in use, but for most companies five to six are all that is needed. Too many explicit measures can obscure the key innovation drivers that make things click. Some companies jump into measuring the wrong things, such as input, when they should be measuring impact. Many choose targets that work in one sector but not necessarily in their own. Others fail to integrate activity metrics to performance goals while still more fail to adopt the targets most applicable to innovation that will drive future growth.

To be a more effective innovator, having the right measures in place is vital. As such, understanding which metrics best fit an organisation's strategy, sector and position within the market is crucial. Measures need to report past performance, track current activities and identify future improvement opprtunities. Furthermore, at the end of the day, these must also be sufficiently transparent so they are understood by, and are useful to, all stakeholders.

SUSTAINABILITY AND INNOVATION

Finally we come to the question of sustainability and its relevance to innovation. Key questions that many are asking include how do companies reconcile the need for sustainable development with creating business success, can both objectives be delivered at the same time, and how do they interrelate? Environmental challenges in particular not only present businesses with obligations, but also with opportunities for innovation. Analysis of the synergies between sustainability and innovation success have been undertaken and these indicate a strong relationship between leadership in innovation and focus on sustainability issues.

In the most widely recognised analysis of sustainability performance, the Dow Jones Sustainability Index (DJSI), companies such as BMW, Philips, Akzo Nobel, Unilever, BT, Intel, TNT, Allianz and Marks & Spencer all came top in their individual sectors in 2007. While all of these companies should be applauded for their achievements, critics who question the long-term business benefit of a sustainable approach ask how this attention to sustainability translates into business growth. Comparing the Innovation Leaders analysis and a recent KPMG International survey gives some clues.

Half of the companies identified in the Innovation Leaders analysis as being amongst the world's top organisations for delivering organic growth also appear in the DJSI with three, Adidas, BMW and Samsung, seen as leaders in both set of analysis. Furthermore a recent KPMG survey analyzing trends in 1600 of world's largest companies in 16 industrialized countries highlighted that over half of the companies in the research see that much of their Corporate Social Responsibility / Sustainability behaviour is motivated by innovation. This suggests that, at one level, when real performance is assessed, attention to the issues that are inferred by being an organisation focused on improving sustainability may be aligned with the ability to deliver significant profitable innovation-driven growth.

The debate on the role of business in society is increasing. Over the past 15 years the global economy has seen unprecedented growth and rising profitability – much of which has come with rising environmental impact. In this context, an increasing number of progressive companies have recognised that it is in their own interest to act responsibly within an increasingly sensitive framework to reduce business risk, secure the long term and safeguard their license to operate. It is noteworthy that a growing selection of companies has started to respond more strategically to a wide range of societal and environmental challenge: From heavy industrial companies through to consumer goods manufacturers, climate change, environmental impact of operations, resource utilisation, carbon footprints and ethical supply chains are all prominent in their strategic priorities. At one extreme, the activities that they are initiating might involve developing technologies that will help to solve or mitigate major environmental problems, while, at the other, it may be more about nurturing new markets in developing countries in a responsible manner. However, one common denominator across many of the leading firms in this area is the ability to create and deliver new innovations by understanding and addressing the challenges present. In either the short term of the long run, many of these translate into competitive advantage and the creation of economic value either for the companies themselves and / or for the communities with which they are working.

Examining the Innovation Leaders and KPMG analysis together supports the view that a relationship between innovation and sustainability exists and that environmental sustainability can be considered a driver for, rather than a barrier to, creating growth and value for organisations and their stakeholders. A closer look at some of the companies that have demonstrated leadership in both shows that, sustainability acts as a catalyst to drive innovation, stimulate alternative ways to develop products, processes and services and also create new opportunities that competitors have not yet recognised.

While many car manufacturers have focused just on flexible fuels and tuning their power systems to run on ethanol and bio-diesel, BMW has also been looking at the use of hydrogen for combustion engines. However, rather than take the same leap into hydrogen fuel cells that thwarted the innovation ambitions of some of its competitors, by being a step too far too early on, the company has instead focused on how it can best take advantage of hydrogen within conventional engines to improve overall effectiveness. At the same time, as it has improved it vehicle's average fuel economy by more than any other firm over the past 15 years, BMW has introduced automatic stop / restart technology that turns of the engine when stationary to reduce emissions but keeps ancillaries such as air conditioning in operation. BMW is also a catalyst for the automotive supply chain to provide alternative materials for body panels that enable safe weight reduction and so reduce energy consumption.

Taking a strategic approach Toyota developed a vision for a 21st century vehicle that complies with sustainability principles by addressing environmental and natural resource issues. One starting point led this automotive giant to conclude that current technologies would never be capable of achieving an ambitious 100% fuel efficiency improvement target. Therefore, building on a number of existing development projects using a variety of hybrid solutions, Toyota was able to radically innovate around the engine concept. This ultimately led to a number of new vehicles under the Lexus and Toyota brand including the globally recognised Prius, cumulative sales of which have now reached nearly 1m units. Whether or not you buy-in to the short term economics of the move, accompanied by parallel developments, this has successfully positioned Toyota well in front of the majority of its competitors when it comes to innovation driven business growth inspired by sustainability.

BMW and Toyota are the only car companies that enjoyed recent increased sales in the US while at the same time cutting CO2 emissions. With Toyota this has mainly been because of the popularity of the Prius and other hybrids, while BMW has continued to make incremental improvements in fuel efficiency in key models – especially in the Mini that now emits 25% less CO2 per mile than the average BMW vehicle. The success of both BMW and Toyota show that their strategies of improving fuel economy have not compromised sales and competitiveness.

Samsung recently came out on top of Greenpeace's green electronics list - ahead of their peers Philips, Sony and LG. Since 2004 and as part of its overall design strategy, the company has been operating an 'eco-design' process that is focused on energy and resource efficiency and environmental impact. Samsung has adopted a 'precautionary principle' by, for example, replacing materials if they show potentially dangerous impacts on human health or the environment. As a result, the environmental credentials of many of Samsung's energy efficient fridges, laptops and LCD displays are now well ahead of most of the competition. As consumers' purchasing decisions are increasingly influenced by these issues, the benefits are clear. Adidas also adopts a precautionary principle similar to Samsung, embedding environmental considerations into their product design processes. The Grün collection is made from

natural recycled materials such as hemp and used denim. In other areas, the company is aiming for completely biodegradable products and is collaborating on product design with Diesel, another firm which is experimenting with biodegradable fibres.

GE's ecomagination is another good example of how a company has made a strategic decision to provide innovative products and solutions which address many of its customers' sustainability issues: All products in the ecomagination portfolio be they in the clean energy, water purification or home appliances spaces, are tracked for both financial contribution to the company's performance and environmental benefit. In addition, in order to be included in the portfolio they have to comply with very specific criteria that address environmental and social impact. The ecomagination portfolio has already contributed well over $10bn to GE's revenues, a figure which the company aims to double by 2010 supported by an additional $700m investment in associated R&D activity. In GE, this strong commitment to sustainability at a strategic level linked to clear financial and environmental targets is visibly driving innovation and helping the company jump ahead of its peers in respective arenas.

Lastly, Wal-Mart's CEO Lee Scott commented on the company's new environmental sustainability strategy: "What I thought was going to be a defensive strategy is turning out to be a precisely the opposite". Wal-Mart, the biggest retailer in the world and a traditional focus for criticism by elements of the media, has drawn up ambitious sustainability targets and is asking suppliers to comply: Those suppliers which saw this trend some time ago, and so proactively developed environmentally-sound products, now have a competitive advantage that they can leverage. This approach is also now to be found in other companies such as for example for Philips' 'Green Flagships' and Unilever's smart packaging solutions, both of which are being emulated by Wal-Mart.

In addition to these established companies, the number of emerging businesses able to identify new spaces related to sustainability is also increasing. Companies such as Greenorder, which helped GE in setting up Ecomagination and Ecosecurities to develop projects to mitigate CO_2 emissions, is growing its businesses as a result of the drive for sustainability. Multiple start-ups in the renewable energy space are exploring new technologies to improve solar panel efficiency, produce fuel from algae and develop more efficient automotive solutions. Furthermore, many entrepreneurs and venture capitalists are seeing new opportunities such as using locally available renewable energy to power small water purification solutions that give more people access to clean water and light after sunset in developing countries.

While these examples highlight intentional focus on innovating in and around the sustainability agenda, a question that remains is how will organisations consciously unlock the often 'hidden' innovation potential around

Sustainability and innovation clearly go hand in hand and so need to be aligned within many companies' strategic priorities

environmental sustainability? On one hand it must be part of the organisation's supporting culture through inclusive thinking but, on the other, it has to be part of the innovation process. This can be quite a challenge, especially for those not fully familiar with the implications and benefits of moving towards an innovation approached focused on more sustainable societal changes, technological development and market dynamics as well as shifts in the political and regulatory landscapes.

There are five key elements to address if an organisation is to successfully embed the environmental sustainability agenda in its innovation process.

1. Organisations have to make a mental leap towards recognising sustainability as a driver rather than a barrier for innovation. This enables them to see opportunities and growth platforms where others fear problems and increased costs.
2. Firms need to understand that sustainability is broader than a company's environmental footprint, providing good working conditions for employees or respecting local communities in which it operates. It is just as much about how their current and future products and services will potentially impact the sustainability requirements of both their customers and ultimately the consumer.
3. As with any insight-driven innovation process, companies must have the capability to draw in and learn from external views from the edge of their arena of focus and then translate them into actions for their current and future business operations, as well as seeing the new roles that an organisation can play across the value chain.
4. Measurable targets are another important area that needs to be addresses to ensure that sustainability is not treated as a 'nice to have' but is core and seen as essential to the processes of creating value in financial as well as environmental terms.
5. Finally, for any company, the change towards more sustainable innovation needs to be inspired by clear leadership, demonstrated by public commitment and driven by internal support for those who undertake the activities and lead programmes that will fuel future growth.

Several companies have shown this can be done and an increasing number of organisations are now following the leaders, taking on board insight in sustainability as a continuous driver in the innovation process at a more strategic level. So, if the ingredients for successful sustainable innovation are clear, the ultimate challenge is to change the mindset and perception of the company around the broad sustainability agenda and the innovation process to collectively drive sustainable growth. Or in the words of Charles Darwin 'It is not the strongest or most intelligent that survive, but the ones that are most responsive to change'.

While the ultimate value to be gained from all of this is still largely assumed rather than validated, the observed momentum is showing that more organisations and entrepreneurs believe it is significant enough, both in financial as well as environmental impact, to influence major strategic choices. As the US becomes more tuned in to the challenges of climate change, China and India align priorities with the global consensus and the economic advantages of a sustainable approach to development become apparent to all, many see that innovation driven by the environmental sustainability agenda will soon become as common as that driven by understanding consumer needs. Sustainability and innovation clearly go hand in hand and need to be aligned within many companies' strategic priorities.

CONCLUSION

innovation

FUTURE INNOVATION

The Innovation Leaders profiles and additional insights provide a view of the current state of the art in innovation. They highlight not only which companies are making the most impact right now, but also give perspectives on the varied strategic and operational approaches being adopted to keep these organisations ahead of the curve. Many of these more recent approaches may soon become new common ingredients in the innovation portfolio. As other companies seek to improve their innovation performance, there is an increasing range of practices that need to be considered and appropriately adopted and adapted.

However, as the Innovation Leaders and their peers seek to further enhance their innovation capabilities, what are the new additional areas that they will be looking to explore? What are the emerging innovation trends that will be potential future sources of differentiation? We see a whole range of potential issues on the horizon that will impact different sectors to lesser or greater extents. Within these, there are two issues that many companies will need to consider:

Extreme Simplicity: The simplicity movement has gained ground as companies have sought to make technology more accessible and consumer devices more intuitive. From Apple's iPod to the shift in Philips' entire corporate focus, simplicity is now become main-stream for many firms. Moving ahead, more radical 'extreme simplicity' is coming to the fore: the $100 One Laptop Per Child is finally making an impact in the developing economies it is designed for, while Tata's new Nano fundamentally rethinks what a mass-market car should be. Both are designed to be extremely simple but deliver the functionality required to meet the core needs of the target market. In addition, several mobile phone companies are launching stripped down products - not just for emerging markets - while some of the world's leading banks are looking at more simplified financial products. Rather than being an afterthought, making products and service extremely simple is destined to be the source of many new breakthrough innovations over the next few years.

Open Open Innovation: Currently, most companies' open innovation programmes are engaging with the usual suspects that populate the ecosystems within which the hosts choose to position themselves. Programmes such as Connect + Develop, Innocentive and backstage.bbc.co.uk have been effective in providing an easy entry point to the host organisations for entrepreneurs, and a means for companies to post problems and solicit solutions. However, while successful to varied degrees, few of these programmes have yet to deliver significant breakthrough opportunities. Going forward, as companies strive for more radical ideas, expect to see broader, more diverse open innovation initiatives that seek to connect the unusual suspects in new ways around common agendas while simultaneously sharing any intellectual property.

As the global innovation landscape accelerates further, these and other strategic and organisational developments will be the drivers of change in the future innovation space. They will be the future sources of higher margins and sustained growth, and so may well be the core of future Innovation Leaders' credentials. Innovation is now a core issue for any growth-focused business and, as such, keeping abreast of the latest approaches, understanding what makes them work and seeing how to adapt them will be a necessity for all. Innovation drives growth and so demands clear focus.

INNOVARO

The Innovation Leaders analysis is undertaken by Innovaro, Europe's leading innovation strategy and insight firm that works with many of the world's major organisations to stimulate, define and deliver innovation. Acknowledged leaders in the field, Innovaro provides insight into the key drivers on, and opportunities for, innovation. The company is recognised as a prime source of innovation expertise and seen as eminent in accessing five areas of insight:

Market insight:	Making sense of the marketplace and clarifying the emerging opportunities
Consumer insight:	Getting under the skin of the consumer to reveal the key influences on decisions
Society insight:	Providing in-depth understanding of how the world we live in is changing
Technology insight:	Highlighting the sources of the emerging technologies that will change our lives
Organisation insight:	Understanding how, why and where leading innovators deliver their potential

Innovaro works at the leading edge of innovation practice, helping clients to address the key issues and create value.

Innovaro's Perspective on Insight

Insight that stimulates, challenges, guides and prioritises corporate activities is an increasingly important factor in delivering successful innovation. Insight drives innovation – so gaining access to the most relevant and up-to-date insight across multiple areas is a capability that many leading companies increasingly need. This is a field that Innovaro has developed as a core arena of expertise. We think of insight as deep-rooted understandings and fresh perspectives about society, markets, technology, organisations and consumers that, when connected in the right way, help to identify and drive commercial ideas that will help to invent and deliver the future.

If insight, ranging from short-term consumer preferences and market shifts right through to long-term changes in societal structures and priorities, is to have value it must be relevant to the needs and culture of the organisation. In a world of Google, many believe that they have easy access to the latest insights. However, as many now recognise, getting secondary information may be easy, but translating this into useful intelligence is not. More than this, having the connections and skills to interrogate data and intelligence, using the most revealing research techniques with lead consumers or consulting with world leaders in respective fields, is a capability that few organisations truly possess.

As Innovaro's clients seek guidance for their innovation decisions, Innovaro has built up a class-leading capability in accessing and interpreting new developments across the five key areas of insight. Alongside the use of the Innovation Leaders analysis, Innovaro Insight brings together an extensive range of knowledge in an accessible form to provide clear and succinct views of today and the future, giving organisations greater clarity on the key elements that can be used to stimulate and steer their innovation activities.

Contact Innovaro:

Innovaro is based in Europe and supports organisations across the globe. Its main offices are in London, Amsterdam, Cambridge and Munich.

London	Amsterdam	Munich	Cambridge
T: +44 (0)207 866 6184	T: +31 (0)20 520 79 46	T: +49 (0)89 99 21 61 01	T: +44 (0) 1223 393 507
london@innovaro.com	amsterdam@innovaro.com	munich@innovaro.com	cambridge@innovaro.com

DERIVATIVES
copyright
ginator or th
publish,

IMAGE COPYRIGHTS

Adidas - © Copyright Adidas AG, 2008

Apple - © Copyright Apple, Inc. 2008

Aviva - © Copyright Aviva plc, 2008

BASF - © Copyright BASF SE, 2008

BMW - © Copyright Bayerische Motoren Werke AG, 2008

Boeing - © Copyright Mark Wagner, 2008

Canon - © Copyright Canon Inc. 2008

Google - © Copyright Google Inc. 2008

Handelsbanken - © Copyright Svenska Handelsbanken AB 2008

H&M - © Copyright Hennes & Mauritz AB 2008

Infosys - © Copyright Infosys Technologies Limited, 2008

LEGO - © Copyright LEGO Holding A/S, 2008

Medtronic - © Copyright Medtronic, Inc. 2008

Microsoft - © Copyright Microsoft Corporation, 2008 Reprinted with permission of Microsoft Corporation

Nokia - © Copyright Nokia Corporation, 2008

NTT DoCoMo - © Copyright NTT DoCoMo, Inc. 2008

PepsiCo - © Copyright PepsiCo, Inc. 2008

Reckitt Benckiser - © Copyright Reckitt Benckiser Group Plc, 2008

Samsung - © Copyright Samsung Electronics Co., Ltd 2008

Shell - © Copyright Royal Dutch Shell plc, 2008

Starwood - © Copyright Starwood Hotels & Resorts Worldwide, Inc. 2008

Tesco - © Copyright Tesco plc, 2008 Courtesy Tesco

UPS - © Copyright United Parcel Service of America, Inc. 2008

Virgin Atlantic - © Copyright Virgin Atlantic Airways, 2008

All other images sourced from istock